Information on the evolution, taxonomy, morphology, anatomy, physiology and genetics of grapevines is scarce and spread thinly in the literature. This book aims to provide for the first time in English a concise but comprehensive overview of the biology and cultivation of the grapevine, accessible to all concerned with viticulture.

After a description of the essential features of viticulture, including a concise history from antiquity to modern times, the authors consider the taxonomy of the grapevine and the evolutionary processes which gave rise to the diversity within the *Vitaceae*. Particular attention is paid to the genera *Vitis* and *Muscadinia*, which are considered a reserve of genetic variation for the improvement of grapevines. A description of the vegetative and reproductive anatomy of the grapevine precedes a full discussion of the developmental and environmental physiology of these fascinating and economically important plants. The concluding chapter considers the potential for genetic improvement of grapevines and includes coverage of the problems encountered, and the methods and strategies employed, in breeding for scions and rootstocks. Special reference to the role of plant biotechnology and tissue culture in the genetic improvement of grapevines is also made.

BIOLOGY OF THE GRAPEVINE

BIOLOGY OF HORTICULTURAL CROPS
Edited by Michael G. Mullins†

Existing texts in horticultural science tend to cover a wide range of topics at a relatively superficial level, while more specific information on individual crop species is dispersed widely in the literature. To address this imbalance, the *Biology of Horticultural Crops* series presents a series of concise texts, each devoted to discussing the biology of an important horticultural crop species in detail. Key topics such as evolution, morphology, anatomy, physiology and genetics are considered for each crop species, with the aim of increasing understanding and providing a sound scientific basis for improvements in commercial crop production. Volumes to be published in the series will cover the grapevine, citrus fruit, bananas, apples and pears, and stone fruit.

BIOLOGY
OF THE GRAPEVINE

Michael G. Mullins†

Department of Viticulture and Enology
University of California, Davis, USA

Alain Bouquet

INRA Station de Recherches Viticoles, France

and

Larry E. Williams

Department of Viticulture and Enology
University of California, Davis, USA
and
Kearney Agricultural Center, Parlier, USA

CAMBRIDGE
UNIVERSITY PRESS

CAMBRIDGE UNIVERSITY PRESS
Cambridge, New York, Melbourne, Madrid, Cape Town, Singapore, São Paulo

Cambridge University Press
The Edinburgh Building, Cambridge CB2 8RU, UK

Published in the United States of America by Cambridge University Press, New York

www.cambridge.org
Information on this title: www.cambridge.org/9780521305075

First published 1992
Eighth printing 2004

A catalogue record for this publication is available from the British Library

Library of Congress Cataloguing in Publication data
Mullins, Michael G.
Biology of the grapevine / Michael G. Mullins, Alain Bouquet,
and Larry E. Williams
p.   cm.
ISBN 0-521-30507-1 (hardback)
1. Grapes.   2. Viticulture.   I. Bouquet, Alain.
II. Williams, Larry Edward.   III. Title.
SB388.M85 1992
634,8–dc20   91-28776   CIP

ISBN 978-0-521-30507-5 hardback

Transferred to digital printing 2007

Contents

Preface

To say that there is no other book like it is to tempt hostile comment from reviewers, but this statement is true, nevertheless. *Biology of the grapevine* is a unique publication.

Information on the evolution, taxonomy, morphology, anatomy, physiology and genetics of grapevines is scarce and is thinly spread in the literature on horticulture and the plant sciences. The grapevine is not a major crop plant of the English-speaking world and the few textbooks available on the biology of the grapevine, and on the fundamentals of its cultivation, are written in French, Italian, German or Spanish. This creates difficulties for English-speaking students, many (if not most) of whom lack the necessary language skills to cope effectively with foreign textbooks. English language journals such as *Vitis* and *American Journal of Enology and Viticulture* are excellent sources of information on current research, but they do not fulfil the role of textbooks.

The aim of this book is to provide an overview of the biology of the grapevine and an introduction to the underlying principles of viticulture. The production of a comprehensive, concise, text-cum-reference book is a challenging task. The treatment of some topics in this book may be too superficial for some readers. Other readers may be disappointed that so many important topics have been omitted. A saving grace is that the book contains extensive references. If researched, these references will greatly augment the text and will provide the reader with good coverage of most, if not all, areas of significance in viticultural science.

Biology of the grapevine is aimed primarily at senior undergraduates in horticultural science, and at beginning graduate students whose research projects are concerned with vines or wines, and it assumes an understanding of elementary plant science. However, many of the chapters are not excessively technical in content and it is hoped that this book will appeal to general readers as well as to specialists, including those associated with the grape and wine industries.

Michael G. Mullins

Davis, California, USA
October 1990

Acknowledgements

The completion of this book has been delayed by changes in authorship, by illness, by the transfer of the senior author in 1987 from the University of Sydney, Australia, to the University of California, Davis, U.S.A., and by his sad death on 13 November 1990. Meanwhile, horticultural science editors at Cambridge University Press have come and gone but the support of the Press has been unwavering. Special thanks are due to Alan Winter, Martin Walters, Katherine Willis, and Maria Murphy. The authors thank B. Bravdo, L.P. Christensen, N. Dokoozlian, M.C. Goffinet, A. Lakso, M.V. McKenry, A. Walker and R. Wample for reviewing and providing helpful comments on various chapters of the book.

Special acknowledgement to Dr Mark A. Matthews, Associate Professor of Viticulture, University of California, Davis, for his contribution of pages 121–40, and to Douglas Fong for his valued assistance with illustrations.

It is a particular pleasure to acknowledge the clerical skills, unfailing good humor and hard work of Mrs Susan Woody, who was responsible for the typing and coordination of the manuscript from 1987 to 1990.

Introduction

The special characteristics of viticulture: Setting the scene for scientific study of the grapevine

A fundamental difference between broad-acre agriculture and horticulture is that innovation in field crops is primarily in the genotype but innovation in fruit-growing, including viticulture, has been mainly at the level of husbandry. This is well illustrated by comparing wheat-growing with the production of grapes. In cereal crops new cultivars are introduced in response to changes in biological or economic factors of production such as epidemics and new technology, and the lead-time for release of an improved or disease-resistant cultivar is about 10 years. By contrast, the cultivars of fruit crops change slowly or not at all. Most of the world's vineyards are planted with traditional grapevines, which have been perpetuated for centuries by vegetative propagation. For example, the Shiraz vine (syn. Syrah, Hermitage) is thought to have originated in Syracuse and to have been introduced into the valley of the Rhône by the Romans. The history of most of the other traditional cultivars is equally long and equally fascinating.

In viticulture the main response to biological constraints or economic change has been to manipulate the existing traditional cultivars by applying progressively higher inputs of husbandry. Included are innovations in standard husbandry (rootstocks, pruning, training), in chemical-based husbandry (fertilizers, pesticides, herbicides, growth regulators), in mechanization (mechanical harvesting and pruning) and in postharvest technology and processing (winemaking). In viticulture we have superimposed twentieth-century technology onto the technology of Rome, a striking illustration of which is the sight of a mechanical harvester straddling the rows of Shiraz or of some other ancient cultivar.

There has been progress in the development of new rootstocks during the past century, but plant breeding has made relatively little impact on

I

viticulture at the level of the scion. The reasons for the persistence of traditional European cultivars of wine grapes, both in their countries of origin and in the New Worlds of North and South America, South Africa and Australasia, are many and involve a complex mixture of plant and human factors.

Technical constraints in grape breeding and vineyard management

Wine production accounts for 80% of the world's grape crop. Emphasis is given here to wine grapes, but table grapes and grapes grown for drying (raisins) are subject to similar circumstances. The breeding of woody perennial fruit plants such as grapevines presents formidable technical difficulties. Grapevines are highly heterozygous outcrossers and they do not breed true from seed. Moreover, the characters that make a good cultivar are polygenic in their inheritance and are controlled by large numbers of genes of minor effect. Few traits of viticultural importance are controlled by single genes with dominant alleles. The traditional grapevine cultivars represent highly subtle gene combinations and these are conserved by vegetative propagation. A high technology has been developed to grow the traditional cultivars and this has become established by custom or by law, particularly in Europe. The wines produced by these cultivars have unique characteristics of style and quality, enjoy a high level of consumer acceptance and are firmly entrenched in the market place.

Pressures for change

Strong pressures for change in methods of production are being applied to all fruit-growing industries, including viticulture, by economic and social forces. The cost of production of modern intensive viticulture is high and includes a significant energy component. Chemical crop protection is not only expensive but also the cause of disquiet to consumers in regard to public health, the wholesomeness of foodstuffs and damage to the environment by toxic residues. This disquiet is being expressed in increasingly stringent environmental protection legislation, a side-effect of which has been a slowing down in the rate of development of new plant protection compounds. Rural industries in most developed countries are becoming more vulnerable to new pests and diseases and are becoming restricted in the ways in which they can deal with existing pests and diseases. All of these factors point to the conclusion that chemical controls must be replaced by genetic resistances, that husbandry must be simplified and that a high priority must be given to plant breeding. Given the technical difficulties, and other disincentives to innovation in viticulture, it is reasonable to question whether these goals are attainable. The advance of knowledge in the plant sciences encourages an optimistic response. This book has been written in the belief that a better appreciation of the biology of grapevines will ease the difficult task of reconciling the opposing forces of tradition and innovation.

Recommended reading

Amerine, M.A. 1964. Wine. *Scient. Am.* **211**: 46–56.

Amerine, M.A. and Singleton, V.L. 1977. *Wine: an introduction.* Second edition. University of California Press, Berkeley. 370 pp.

Bouquet, A. 1982. Origine et évolution de l'encépagement français à travers les siècles. *Progr. Agric. Vitic.* **99**: 110–21.

Winkler, A J., Cook, J.A., Kliewer, W.M. and Lider, L.A. 1974. *General viticulture.* University of California Press, Berkeley. 710 pp.

I

The growing of grapes

The history of viticulture

INTRODUCTION

The growing of grapes and the making of wine have a prominent place in the history of Western civilization. The ancients gave an importance to wine which greatly exceeded its role as a beverage. They regarded wine as a gift from the Gods and possessed of mystical significance. The Egyptians attributed the gift of wine to Osiris, the Greeks to Dionysos and the Romans to Bacchus. The Old Testament contains many references to wine. In Genesis it is written that after the Flood 'Noah began to be an husbandman and he planted a vineyard and drank of the wine' (Fig. 1.1). In Christianity the need for sacramental wine led to an association between grape growing and the Church which has flourished for many centuries. The traditions of viticulture and the mythology of wine have given the grapevine a privileged position among cultivated plants.

It is thought that cultivation of the grapevine began during the Neolithic era (6000–5000 BC) along the eastern shores of the Black Sea in the region known as Transcaucasia, but archaeological finds of grape seeds indicate that *Vitis vinifera* L., or its progenitor, *Vitis sylvestris*, was distributed throughout much of Europe during the Atlantic and Sub-Boreal palaeoclimatic periods (7500–2500 years ago). Grapevines occurred as far north as Belgium, and recent archaeological work in Spain has uncovered grape seeds which, according to radiocarbon dating, are aged between 4350 and 3950 years (Walker, 1985). It is likely that prehistoric people ate the berries of the wild vines that climbed on the forest trees. The discovery of wine was probably accidental. The fruit of the grapevine provides an ideal substrate for fermentation and the surface of the grape berry is a favorable habitat for yeasts. It is probable that wine was the unexpected (if inevitable) result of the storage

4

Fig. 1.1. Noah and the grapevine. Mosaic, Basilica of St Mark, Venice

Fig. 1.2. Grape growing and winemaking in Sixth Dynasty Egypt (Tomb of Mererou-Ka). Date: *c*.2345–*c*.2181 BC. Modified from Dage and Aribaud (1932)

of grapes during the winter months. With the development of village settlements it is likely that the best of the wild vines from the forest were brought into cultivation, and thus began the history of viticulture (grape growing) and enology (the art and science of winemaking).[1]

By 4000 BC, grape growing and the making of wine extended from Transcaucasia to Asia Minor, through the Fertile Crescent and into the Nile delta. Viticulture is illustrated in mosaics of the Fourth Dynasty in Egypt (2440 BC); the technology of grape growing and winemaking was well developed by 1400 BC (Fig. 1.2). Laws on the wine trade, and on wine consumption, were enacted by King Hammurabi of Babylon about 1700 BC. It is said that grape culture first appeared in China about 2000 BC but was prohibited by the Emperor. It is more likely that cultivated grapes were brought to China from Asia Minor during the Han Dynasty (second century BC) (Huang, 1980).

THE GRAPEVINE IN EUROPE

The Greeks of Homer were wine drinkers but the origins of viticulture in Greece are uncertain. About 3000 BC the expansion of the Hittites in Anatolia led to significant migrations towards the west. These refugees occupied Crete and the Aegean islands and gave rise to the famous Minoan civilization (2200–1400 BC), which subsequently had considerable influence on the culture and commerce of ancient Greece. Grape

---

[1] The term, *viticulture* is of Latin derivation (*Vitis*, grapevine) but enology (or oenology) comes from *oenos*, the Greek word for wine. The Greek for grapevine is *ampelos*; this is reflected in the French term for the study of grapevines, *ampélologie*, and in a former name for the grapevine family, *Ampelidaceae*. The family is now known as *Vitaceae*.

growing and winemaking may have been introduced into Crete, and thence to the Pelponnesus, during this period but the possibility of a direct introduction into the Balkan Peninsula through the Bosporus cannot be excluded. In the Iliad, Homer says that the inhabitants of Thrace supplied the Greeks with wine during the Trojan War (1300 BC) and Pliny reports in his *Natural History* that Eumolpus, a Greek from Thrace, was responsible for teaching the Romans the cultivation of grapevines.

During the first millennium BC the Phoenicians and Greeks extended their influence to the western Mediterranean; they introduced viticulture into North Africa (Carthage), Sicily, southern Italy, Spain and France. The first vineyards in the south of France were planted by a settlement of Greeks at Massilia (Marseille) about 500 BC (Levadoux, 1953), but viticulture did not become widespread until the establishment of the Roman province of Narbonnaise in the first century BC. The first Greek settlers in southern Italy were so impressed by the ability of vines to flourish in this region that they called their new home Oenotria: the Wineland. Later, the growing of grapes spread throughout Italy and the importance of viticulture to the economy of Rome, and in Roman culture, is attested to by the writings of Cato, Varro, Pliny, Vergil and Columella (Billiard, 1913).

Under the influence of Rome, grape growing spread throughout the valley of the Rhine and into Germany. In AD 92 the Emperor Domitian issued an edict to limit the growth of provincial viticulture and to protect the export of wine from Italy. Nevertheless, by AD 300 the growing of grapes and the making of wine extended throughout Europe from the shores of the Atlantic to the valley of the Danube.

With the fall of the Roman Empire the wine trade was disrupted and commercial grape growing went into decline. During the first part of the Middle Ages (AD 500–1000) the custodians of viticulture and the art of winemaking were the monasteries. The spread of Christianity to northern Europe, and the need for wine for sacramental purposes, led to the establishment of a new international trade in wine. This trade soon outgrew its religious association and wine drinking was a firmly established social custom over most of Europe by the end of the Middle Ages (Enjalbert, 1975).

European viticulture grew steadily from the sixteenth to the nineteenth centuries despite a series of calamities: the Thirty Years' War (1618–1648), which ruined the vineyards of the Palatinate (Rheinpfalz), the frost of 1709, which severely damaged the northernmost vineyards of France and Germany and the phylloxera plague, which appeared in France in 1868 and then spread progressively through all the grape-growing regions of Europe. The growth of viticulture went hand-inglove with the growth of an extensive international trade in wine in which the British and the Dutch were major customers.

THE GRAPEVINE UNDER ISLAM

By AD 600 the consumption of wine had been prohibited under Islamic law and this promoted the cultivation of table grapes in the Middle East and North Africa. It also led to the introduction of table grape cultivars into Spain, whence they spread to France, Italy and the New World. The growing of table grapes was also introduced into the Balkans during the Ottoman occupation (15th–19th centuries) but the Turks, unlike the Arabs, permitted their Christian subjects to cultivate wine grapes.

VITICULTURE IN THE NEW WORLD

In 1525 Cortez ordered the planting of grapevines in Mexico and by 1550 the growing of grapes had spread to Peru, Chile and Argentina. Don Pedro del Castillo founded Mendoza in 1556 and established the Argentinian wine industry with grapevines brought from Chile. Colonial viticulture was kept in check by legislation to protect wine exports from Spain. For three centuries, only the Catholic missions were permitted to grow grapes and to make wines, ostensibly for sacramental purposes.

The first grapevine to be grown on the west coast of North America was planted in 1697 by a Jesuit priest, Father Juan Ugarte, at the Mission San Francisco Xavier in what is now Lower California. From 1769 to 1810 numerous missions were founded in California and the cultivation of grapes was extended northwards to Los Angeles and San Francisco.

The east coast of North America was visited by Viking adventurers, who named it Vineland in recognition of the many types of wild grapevines which they found. Later, colonists brought the European grape from their homelands, but it did not thrive in eastern America owing to lack of winter hardiness and susceptibility to pests such as phylloxera and diseases such as downy mildew (these pests and diseases were exported back to Europe in the nineteenth century, with disastrous consequences). Meanwhile, viticulture along the eastern seaboard was based on native American species of *Vitis*, for example, *V. labrusca*, or on interspecific hybrids between North American species and *V. vinifera* (Hedrick, 1908).

The gaining of independence by the countries of South America in the first half of the nineteenth century led to a large expansion of viticulture, particularly in Chile and Argentina. In Uruguay and southern Brazil the wine industry was founded by Italian migrants. Immigration also played a large part in the establishment of commercial viticulture in California during the second half of the nineteenth century. Many experienced vignerons fled to California following the failure of revolutions in Europe in 1848 and 1849. One of these refugees was a Hungarian aristocrat, Agoston Haraszthy, who is now recognized as a founding father of Californian viticulture (Carosso, 1951). The wine industry of the west coast of the United States suffered a major setback during the era of Prohibition (1920–1933).

VITICULTURE IN AFRICA AND AUSTRALASIA

The first grapevines in South Africa were planted in 1616 by Dutch settlers in the Cape of Good Hope. Grape growing flourished and the arrival of French Huguenots provided extra skills in wine production. The sweet dessert wines of the Cape found a ready market in England during the Napoleonic Wars. In North Africa the colonization of Algeria by the French in 1830 led to the restoration of wine growing after its suppression for several centuries under Islam. Later, many vignerons who had been ruined by the phylloxera plague in metropolitan France migrated to Algeria where they developed a large grape-growing industry based primarily on the export of bulk wines.

Viticulture in Australia began in 1788 with the foundation of a penal colony at Botany Bay. Attempts to grow grapes in the vicinity of Sydney were unsuccessful owing to the humid climate, but viticulture became established in the Hunter Valley of New South Wales by 1820 (Busby, 1825). In the 1850s German Lutherans founded the now important wine industry of the Barossa Valley in South Australia. At the end of the nineteenth century, irrigation settlements were established along the Murray River in South Australia and Victoria and the growing of grapes for drying became the predominant industry of this region. In New Zealand, grapes were planted in the North Island early in the nineteenth century by French settlers and by religious orders.

World grape and wine industries

Grapes have many uses; fresh fruits, dried fruit (raisins), fresh grape juice, concentrated grape juice (the 'pecmez' of Turkey), table wines, sparkling wines, champagne, fortified wines (sherry, port and other aperitif wines) and distilled liquors derived from wine (cognac, armagnac, brandy). In addition there are several industrial uses of grapes and grape products including grape-seed oil, anthocyanin pigments and ethanol production, but more than 80% of the world's grape crop is used for wine production.

The grapevine is the world's most widely grown fruit plant; it is cultivated on all continents except Antarctica. In 1988, grape growing accounted for 9.0 million hectares.[2] Europe has a dominant position in the grape and wine industries; it accounts for 70% of world vineyard area, 80% of world wine production, 55% of table grapes and 30% of raisin production. Three countries, Italy, Spain and France, each have more than one million hectares of grapevines. It is noteworthy that nearly one quarter of all the world's wine is made in Italy. The United States, with 300 000 hectares of grapevines, is the eighth largest grower. Grape growing in Australia and New Zealand accounts for less than 1% by area of world viticulture.

WINE PRODUCTION

In 1988 world production of wine was 27 500 megaliters. Three countries, Italy, France and Spain, produce more than 50% of the world's wine; European countries (together with the Soviet Union) account for 80% of world production. There has been little change in the total area devoted to grape growing during the past 20 years. A decline in viticulture in Spain and France, and in Algeria following independence from France, has been balanced by the growth of the wine industry in the USSR. Meanwhile, world production of wine increased by 35% from 1951. This increase in wine production from essentially the same area of vineyards can be attributed to improvements in mechanization and plant protection, and to the availability of pathogen-free stock. Increased production has not been matched by increased demand; most markets are now oversupplied. The main consumers of wine are the wine-producing countries, but there has been a marked drop in consumption in recent years in many of the traditional wine-drinking nations. In France, for example, consumption per capita was 127 l in 1963 and 74 l in 1988 – a reduction of nearly 60%. This gap has been offset to some extent by an increase in wine consumption in northern European countries, the USA and Canada, but despite the growing international trade in wine, only 14% of world wine production is exported.

[2] Statistics on world viticulture are published annually in the *Bulletin de l'OIV*, published by the Office International de la Vigne et du Vin (Paris). The OIV is an intergovernmental organization, founded in 1924, with representation from 33 nations. OIV is concerned primarily with regulatory and legislative aspects of grape and wine industries.

In some countries large quantities of wine are distilled to produce high-proof spirits (cognac, armagnac and brandy) and fortifying spirit for the making of sherry, port or other aperitif beverages. Distillation is also used as a means of controlling the market for wine.

PRODUCTION OF TABLE GRAPES

Statistics on table grape production are less complete than those for wine grapes but the estimated annual crop was 7 Mt for each of the years 1976–88. Table grapes account for less than 12% of the total production of grapes. The ten largest producing countries are Italy (17%), USSR (14%), Turkey (12%), Spain (6%), USA (5%), Bulgaria (5%), Japan (5%), Greece (4%), Brazil (3%) and France (3%). As with wine, table grapes are consumed primarily in the producing countries. A consequence of the perishability of the fruit and of the high cost of transportation is that only 14% of total production is exported. Annual per capita consumption of table grapes is low and does not exceed 10 kg in most producing countries. In Europe and North America table grapes represent less than 5% of annual per capita consumption of fresh fruit.

PRODUCTION OF RAISINS

The term 'raisin' will be used here as the generic term for dried grapes, but the various forms of dried fruit have a variety of names. In the past, 'raisins' were dried seeded grapes, usually Muscat of Alexandria (syn. Muscat Gordo); however, today 'raisins' are considered to be the dried fruit of the seedless cultivar Thompson Seedless (syn. Sultanina). 'Sultanas' are the dried fruits of the cultivar Sultana, and 'currants' are the dried fruits of Zante Currant (syn. Black Corinth). The history of raisin production is, perhaps, as long as the history of wine making. Aristotle (360 BC) referred to the seedless character of the Black Corinth grape, and legend has it that Hannibal fed his troops with raisins during the crossing of the Alps (218 BC). The Ancients used a mixture of wood-ash and olive oil to speed the drying of grapes; the modern 'fruit-dips' that are used to produce 'golden' raisins are of basically similar composition. In this process, freshly picked Thompson Seedless grapes are dipped in proprietary solutions, which contain potassium carbonate and vegetable oil. This treatment disrupts the epicuticular wax or bloom on the surface of the berry and leads to an accelerated loss of water. Fruit treated in this way retains the golden color of the fresh fruit. Water loss in untreated sun-dried Thompson Seedless grapes is much slower than in treated fruit and the berries develop a dark brown color as a result of polyphenol oxidation and non-enzymic browning reactions.

In 1987 worldwide production of raisins was approximately 646 700 tonnes; this corresponds to 25.9 million tonnes of fresh grapes (4 tonnes of fresh grapes produces 1 tonne of raisins). Raisin production is more uniformly distributed among the continents than wine or table grape production: there is one major producer in each of the continents except Asia.

Raisin production is found only between the latitudes of 30° N and 39° N in the Northern Hemisphere and between 28° S and 36° S in the Southern Hemisphere. This is because the two cultivars best suited to raisin production, Zante Currant and Thompson Seedless, require high temperatures for adequate inflorescence formation and for high yields. Moreover, the natural sun-drying of raisins, which is the predominant method of fruit processing, requires hot dry weather during the post-harvest period.

The climatic limits to grape growing

Vitis vinifera L. is a temperate-climate species, which cannot withstand extreme winter cold and which requires warm hot summers for the maturation of its fruits. On a broad scale, the main areas of viticulture are situated between the latitudes of 30° N and 50° N and between 30° S and 40° S. These latitudes approximate to the 10 °C and 20 °C isotherms (Fig. 1.3). Within these zones there are climatic variations caused by mountains, large masses of land or water and ocean currents which greatly affect the distribution of vineyards in the different continents. In Peru, for example, grapes are grown between 12° S and 15° S but the vineyards are in coastal areas, which are subject to the cooling influence of the Humboldt Current. Similarly, the Gulf Stream has an ameliorating effect on the climate of western Europe and grapes are grown in Germany (50–51° N) on the south- and west-facing slopes of the valleys of the Rhine and Moselle. The Southern Hemisphere is cooler than the Northern Hemisphere because of its greater surface of water; the southern limit of grape culture in Chile and New Zealand is generally at latitude 40° S. The world's most southerly plantings of grapevines are in the South Island of New Zealand, close to the 45th parallel. Towards the equator the limiting factors to viticulture are the extreme heat, inadequate winter chilling and lack of water. In the humid tropics Vitis vinifera L. behaves as an evergreen but it can be induced to produce satisfactory crops of fruit with careful management. Table grape production is a small but expanding industry in several tropical countries.

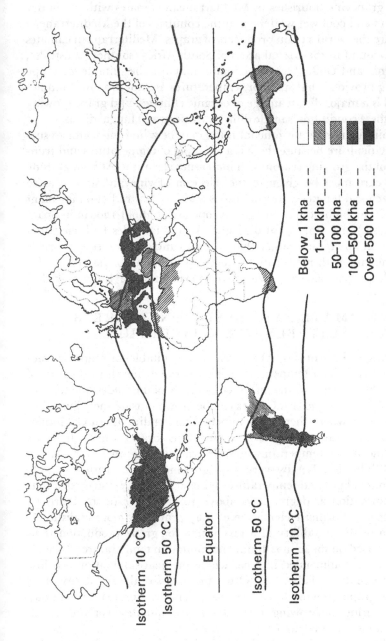

Below 1 kha --- ----
1–50 kha --- ----
50–100 kha --- ----
100–500 kha --- ----
Over 500 kha --- ----

Isotherm 10 °C
Isotherm 50 °C
Equator
Isotherm 50 °C
Isotherm 10 °C

Fig. 1.3. Distribution of wine grape plantings in the world. From Amerine and Joslyn (1970). Reproduced with permission. (Copyright © 1970 The Regents of the University of California.)

The grapevine flourishes in Mediterranean climates with warm dry summers and cool wet winters, and the countries of the Mediterranean basin are the world's main producers of grapes. Mediterranean climates are also found in the coastal areas of South Africa, southern Australia, California and Chile. Oceanic climates may also be suitable for grape growing provided that summer temperatures are not too low. Summer rainfall is a major disadvantage of oceanic climates, and grape growing under these conditions is made difficult by pests and fungal diseases.

Within the temperate zone, altitude may be a limiting factor to successful viticulture because there is a lowering of temperatures and truncation of the growing season with increasing elevation. At high altitude special care must be given to the selection of vineyard sites so as to optimize local temperature conditions and incident radiation for plant growth. The highest vineyards in Europe are at about 1200 m in Spain (Andalusia) and Italy (Val d'Aosta). At low latitudes high elevation may be an advantage rather than a disadvantage because it provides environments with more equable temperatures. In Bolivia (18° S), for example, grapevines are grown above 3000 m but not at lower altitudes.

CLIMATE AND SELECTION OF AREAS SUITABLE FOR VITICULTURE

According to Prescott (1965) a given area is suitable for grape production if (i) the mean temperature of the warmest month is in excess of 18.9 °C (66 °F) and (ii) the mean temperature of the coldest month is in excess of -1.1 °C (30 °F). Grapes can be grown outside these limits but only in areas where there are compensating conditions. In California, for example, grapes are grown in numerous locations near the coast where the mean temperature for the warmest month never reaches 18.9 °C. This is offset by very long growing seasons with favorable conditions of light and temperature. Winkler et al. (1974) suggested that heat summation as degree days above 10 °C for the period April 1– October 31 (Northern Hemisphere) may provide a more realistic indication of the suitability of a given area for grape production than criteria based on mean maximum and minimum temperatures. Viticulture is not recommended in areas where the heat summation is below 1700 degree days (Table 1.1). Other indicators of the suitability of an area for grape growing have been based on heat summation and day lengths during the growing season and heat summation combined with summations of rainfall and hours of sunshine.

Finally, it is worth noting that Vitis vinifera L. is a highly adaptable species and that grapevines can be grown in a very wide range of envi-

Table 1.1. *Environments of grape growing regions*

Heat summations (degree days above 50 °F) for the periods April 1–October 31 (Northern Hemisphere) and October 1–April 30 (Southern Hemisphere).

Meteorological station	Heat summation	Meteorological station	Heat summation
Geisenheim (Germany)	1790	Napa (California, USA)	2280
Reims (France)	1820	Asti (Italy)	2930
Coonawarra (Australia)	2170	Bucharest (Rumania)	2960
Bordeaux (France)	2300	Mendoza (Argentina)	3640
Geneva (New York, USA)	2400	Lodi (California, USA)	3270
Budapest (Hungary)	2570	Sydney (Australia)	3740
Odessa (USSR)	2580	Palermo (Italy)	4100
Santiago (Chile)	2710	Shiraz (Iran)	4390
Melbourne (Australia)	2750	Algiers (Algeria)	5200

Winkler *et al.* (1974).

ronments. Wine grapes are cultivated in the deserts of California and Australia, table grapes are grown in the mountains of Indonesia and at sea level in Thailand, and grapes are grown for winemaking in eastern England. In each case the limitations imposed by climate are countered by inputs of husbandry or technology. These inputs are irrigation in California and Australia, labor for plant manipulation in Indonesia and Thailand, and addition of sugar to the juice for winemaking in England. It should be understood that climate is only one factor in the complex of factors which determines the commercial success of a viticultural enterprise.

Recommended reading

Amerine, M.A. and Singleton, V.L. 1977. *Wine: an introduction.* Second edition. University of California Press, Berkeley. 370 pp.
Winkler, A.J., Cook, J.A., Kliewer, W.M. and Lider, L.A. 1974. *General viticulture.* Second edition. University of California Press, Berkeley. 710 pp.

Literature cited

Amerine, M.A. and Joslyn, M.A. 1970. *Table wines. The technology of their production.* Second edition. University of California Press, Berkeley. 997 pp.
Billiard, R. 1913. *La vigne dans l'antiquité.* Larduchet, Lyon. 560 pp.
Busby, J. 1825. *A treatise on the culture of the vine, and the art of making wine.* R. Howe, Government Printer, Sydney. Facsimile reprint, 1979. The David Ell Press, Hunter's Hill, NSW, Australia. 270 pp.

Carosso, V.P. 1951. *The California wine industry, 1830–1895: a study of the formative years.* University of California Press, Berkeley.

Dage, R. and Aribaud, A. 1932. *Le vin sous les Pharaons.* A. Delayance, La Charité-sur-Loir (Nièvre), 19 pp.

Enjalbert, H. 1975. *Histoire de la vigne et du vin. L'avénement de la qualité.* Bordas, Paris. 208 pp.

Hedrick, U.P. 1908. *The grapes of New York.* (115th Ann. Rept. Dept. Agr. NY.) J.B. Lyon, State Printer, Albany. 564 pp.

Huang, H.B. 1980. Viticulture in China. *HortScience* **15**: 461–466.

Levadoux, L. 1953. De l'origine de la vigne dans les Gaules. *Progr. Agric. Vitic.* **70**: 118–122.

Prescott, J.A. 1965. The climatology of the vine: The cool limits of cultivation. *Trans. Roy. Soc. South Aust.* **89**: 5–23.

Walker, M.J. 1985. 5000 años de viticulture en Espana. *Riv. Arqueol.* **6**: 44–47.

2

The grapevine and its relatives

The family *Vitaceae*

Several names have been proposed for the grapevine family, including *Ampelidaceae* LOWE 1868 and *Ampelideae* KUNTH 1821, the latter being favored by Planchon (1887), but the accepted name under the International Code of Botanical Nomenclature is *Vitaceae*. Grapevines and their relatives are dicotyledonous Angiosperms; *Vitaceae*, together with the families *Rhamnaceae* and *Leeaceae*, comprise the Order Rhamnales, one of the seven orders of the Phylum Terebinthales–Rubiales (Chadefaud and Emberger, 1960).

Vitaceae are mostly woody or herbaceous lianas (tree-climbing plants) or shrubs with liana-like stems and the family is primarily intertropical in its distribution. There are a few exceptions to the climbing habit, but these represent secondary adaptations to desert environments. Morphologically, *Vitaceae* are characterized by the occurrence of tendrils and inflorescences opposite to leaves. This arrangement is difficult to interpret; it is possible that the ancestors of *Vitaceae* were non-climbers. Sussenguth (1953) proposed that leaf-opposed tendrils arose as an adaptation to tropical forest environments, and it may be significant that the genus *Leea*, which was formerly included in *Vitaceae*, contains primitive bushy vines that lack tendrils.

The systematics of *Vitaceae* is based on the classification of Planchon (1887) in which the family comprises 10 genera (Table 2.1). Later revisions were concerned with the addition of two monospecific genera, *Acareosperma* GAGNEPAIN and *Pterocissus* URB. and EK, and with the separation from *Cissus* of the genera *Cayratia* and *Cyphostemma*. At present, the most comprehensive treatment of *Vitaceae* is that of Galet (1967). Another change to the classification of Planchon was the proposal of Small (1903) that the Sections *Euvitis* and *Muscadinia* be elevated to generic level. This proposal, although well supported by

Table 2.1. *The genera of Vitaceae according to PLANCHON, 1887*

I. *Vitis* (Tournef) L.
  Section 1: *Euvitis* PLANCH.
  Section 2: *Muscadinia* PLANCH.
II. *Ampelocissus* PLANCH.
III. *Pterisanthes* BLUME.
IV. *Clematicissus* PLANCH.
V. *Tetrastigma* (Miq.) PLANCH.
	VI. *Landukia* PLANCH.
	VII. *Parthenocissus* PLANCH.
	VIII. *Ampelopsis* MICH.
	IX. *Rhoicissus* PLANCH.
	X. *Cissus* L.
	  Section 1: *Eucissus* PLANCH.
	  Section 2: *Cayratia* JUSS.
	  Section 3: *Cissus* PLANCH.

morphological, anatomical, genetic and cytogenetic evidence (Olmo, 1978; Bouquet, 1980), has not yet achieved general acceptance.

THE GENERA OF *VITACEAE*

Studies on the cytotaxonomy of *Vitaceae* have been made by Shetty (1959) and Lavie (1970). The genera *Vitis, Muscadinia, Ampelocissus, Parthenocissus, Landukia, Ampelopsis, Clematicissus* and *Rhoicissus* belong to a homogeneous group of mostly woody plants which are characterized by pentamerous flowers and by karyotypes of $2n = 40$. An exception is *Vitis*, in which $2n = 38$. In all cases the chromosomes are small and the genera may be of polyploid origin.

Species of *Ampelocissus* are found throughout the tropics and some of them have edible berries. *Parthenocissus* and *Ampelopsis* are plants of the temperate zone and are found mostly in Asia. Exceptions are *Parthenocissus quinquefolia* (Virginia Creeper) and three species of *Ampelopsis*, which are native to North America and which are ornamental climbing plants. The monospecific genus, *Landukia*, is similar to *Parthenocissus*. The genera *Clematicissus* and *Rhoicissus* contain several species of importance in ornamental horticulture.

The genus *Cissus*, formerly *Cissus* sect. *Eucissus* PLANCH., contains more than 350 species and it is distributed throughout the tropics. *Cissus* is a heterogeneous genus with both herbaceous and woody plants with tetramerous flowers. Some species of *Cissus* are important as ornamentals, notably *Cissus antarctica* (Kangaroo Vine), an Australian species which is widely grown as an indoor plant. Cytological studies suggest that the basic karyotype of *Cissus* is $2n = 24$.

The genera *Tetrastigma* and *Cyphostemma* have morphological and anatomical similarities and a common karyotype ($2n = 22$). There are

many cases of polyploidy within each genus. *Tetrastigma* occurs in Asia
and *Cyphostemma* is found in Africa, but the two genera appear to be
closely related. *Cayratia* are found mostly in Asia and are very heterogeneous in morphology and karyotype. A lesser known Asiatic genus,
Pterisanthes, is found in Indonesia.

EVOLUTION OF *VITACEAE*

From studies on the karyotypes and geographical distributions of the
different genera it was postulated by Lavie (1970) that the center of
origin of *Vitaceae* was Asia and that the family evolved from an unknown
ancestor which was similar to *Cissus*. Differentiation of genera is thought
to have occurred during the Mesozoic era before the separation of the
continents. In this hypothesis, the genera *Vitis*, *Ampelopsis* and *Parthenocissus* became established in the future North America and Eurasia,
and the genera *Cissus* and *Ampelocissus* became established in the future
Africa, South America and Australia.

The genus *Vitis*

CLASSIFICATION

The genus *Vitis*, formerly *Vitis* sect. *Euvitis* PLANCH., contains about 60
species. These are found mainly in the temperate zones of the Northern
Hemisphere and they are distributed almost equally between America
and Asia. Only one species, *Vitis vinifera* L., originated in Eurasia; it has
been spread throughout the world by man.

The systematics of *Vitis* has been a subject of controversy for more
than a century. Numerous botanists have classified and provided names
for the grapevines found around the world, and there are many more
names in the literature than there are genuine species. Some order was
brought to this chaos by Galet (1967) who clarified the synonymy
among more than 3000 taxa, and Rogers and Rogers (1978) have
applied numerical taxonomic methods to the identification of North
American species. The determination of the number of 'real' species
of grapes, and of their proper names, has considerable significance for
plant improvement. Many false starts have been made in grape breeding because the correct names and classification of the parents were
unknown. Another requirement for precision in the naming of *Vitis* species is related to the conservation of genetic resources. Wild grapevines,
like the wild relatives or progenitors of many other crop plants, are

being endangered by human population growth and by the expansion of agriculture and industry. A high priority needs to be given to the collection and preservation of *Vitis* germplasm and the usefulness of this enterprise is dependent upon accurate identification.

The first classification of *Vitis* was that of the French taxonomist, Planchon (1887), who placed the American and Asiatic species in separate series. This classification was followed by later French ampelographers such as Foex (1895), Ravaz (1902), Galet (1967) and Levadoux (1968). The earlier of these workers concentrated their interest on the relatively small number of species which were useful germplasm in the fight against phylloxera, the predominant problem of French viticulture in the nineteenth century. Consequently, they tended to give little attention to the species that were without practical interest and their view of the genus *Vitis* was somewhat restricted.

In the USA, extensive work on grape systematics was done by Munson (1909), who was an experienced viticulturist as well as a botanist, but the most complete classification of North American species of *Vitis* was produced by Bailey (1934).

So far, the most comprehensive treatment of the genus is that of Galet (1967) who listed 59 species grouped into 11 series. Included are 25 grape species from Asia, but the standing of many of these Asian *Vitis* is still controversial. The classification given in Table 2.2 includes 43 of the better known species of *Vitis*. The revival of viticulture in China (Huang, 1980) is likely to stimulate interest in genetic resources, and in the taxonomy of Asiatic *Vitis*, and further information on this germplasm may soon be available.

SYSTEMATICS OF *VITIS*

The systematics of *Vitis* is a particularly difficult area of taxonomy; the validity of any classification is still problematical. In the past the definition of a species was based primarily on comparative morphology with support from environmental characteristics (habitat, soils, climate) and geographical distribution. Evidence for natural hybridization was based on the phenotypic resemblance of the supposed hybrid with two or more sympatric species. Given the extreme morphological variation among and within *Vitis* species, and the genetic variability of hybrid populations, it is not surprising that there has been a lack of unanimity in identifying ancestral species. Nevertheless, use of these standard methods has enabled some long-standing errors to be corrected. The cultivar Concord, which is of great importance in New York State, was once regarded as pure *Vitis labrusca* but is now thought to be a natural hybrid

Table 2.2. *Classification of* Vitis *species (partial listing)*

Series	Species and synonym	Origin	Series	Species and synonym	Origin
I. Candicansae	V. candicans	North America (East)	VI. Continued	V. gigas	North America (East)
	V. champinii	North America (East)		V. rufotomentosa	North America (East)
	V. doaniana	North America (East)		V. bourquina	North America (East)
	V. simpsonii = V. smalliana	North America (East)	VII. Cordifoliae	V. cordifolia	North America (East)
	V. coriacea = V. shuttleworthii	North America (East)		V. rubra = V. palmata	North America (East)
II. Labruscae	V. labrusca	North America (East)		V. monticola	North America (East)
	V. coignetiae	Asia		V. illex	North America (East)
III. Caribaeae	V. caribaea = V. tiliaefolia	North America (South)		V. helleri	North America (East)
	V. blancoii	North America (East)	VIII. Flexuosae	V. flexuosa	Asia
	V. lanata	Asia		V. thunbergii	Asia
IV. Arizonae	V. arizonica	North America (West)		V. betulifolia	Asia
	V. californica	North America (West)		V. reticulata	Asia
	V. girdiana	North America (West)		V. amurensis	Asia
	V. treleasei	North America (West)		V. piasekii	Asia
V. Cinereae	V. cinerea	North America (East)		V. embergeri	Asia
	V. berlandieri	North America (East)		V. pentagona	Asia
	V. baileyana	North America (East)	IX. Spinosae	V. armata	Asia
	V. bourgeana	North America (South)		V. davidii	Asia
VI. Aestivalae	V. aestivalis	North America (East)		V. romaneti	Asia
	V. linecumii	North America (East)	X. Ripariae	V. riparia = V. vulpina	North America (East)
	V. bicolor = V. argentifolia	North America (East)		V. rupestris	North America (East)
			XI. Viniferae	V. vinifera	West Asia and Middle East

Based on Galet (1967).

of *V. labrusca* and *V. vinifera* (Tukey, 1966). The cultigen name, *Vitis* ×
labruscana L. H. Bailey, is used to designate American grape cultivars
having *labrusca* parentage (Cahoon, 1986).

CYTOTAXONOMY AND CHEMOTAXONOMY

Cytotaxonomic methods have been useful in research on the *Vitaceae* at
large, but information on chromosome numbers within the genus *Vitis* is
inadequate for the determination of species or groups of species. The
application of chemotaxonomy to *Vitis* has been much more successful.
Ribéreau-Gayon (1959) showed that the anthocyanins of *V. vinifera* are
monoglucosides and that those of most North American species are
diglucosides. The diglucoside character is dominant, and these compounds occur in the wine of interspecific hybrids. Accordingly, a rapid
test was developed for adulteration of good quality wine from *vinifera*
grapes by poorer quality wine from high-yielding hybrids. In taxonomic
work by Olmo (1980), spontaneous hybrids of *V. californica* and *V.
vinifera*, which closely resembled *V. vinifera* and had hermaphrodite
flowers and large berries, were proved to be hybrid by the presence of
anthocyanin diglucosides. Differences in the occurrence of amino acids
in leaf tissue also provide a means of discriminating among groups of
species. The species that compose series V and VI of Galet (Table 2.2),
V. cinerea, *V. berlandieri*, *V. aestivalis* and *V. rufotomentosa*, all contain
hydroxyproline but not homoserine (Kliewer *et al.*, 1966). Isozyme analysis (Schaefer, 1971; Wolfe, 1976) and serological analysis of pollen proteins have also been employed in research on *Vitis* systematics (Samaan
and Wallace, 1981). In the future, it is likely that much greater use will
be made of techniques in genetics, both conventional and molecular,
and in biochemistry for the identification and classification of *Vitis*
species.

THE EVOLUTION OF SPECIES IN THE
GENUS *VITIS*

There are many references to prints of grapevine leaves in rocks from
Tertiary deposits. Some of these fossils have been attributed to archaic
genera of *Vitaceae*, such as *Cissites* HERR or *Paleovitis* RED and
CHANDL. Others have been attributed to a fossil species of *Vitis*,
V. sezannensis. In fact, the identity of these fossilized leaf-prints is
extremely doubtful and they could equally well be attributed to plants
other than grapevines. Only the fossilized prints of seeds are acceptable
as evidence for the existence of *Vitis*. According to Kirchmeimer (1939)

two groups of fossilized seeds assignable to *Vitis* have been found in the Tertiary sediments of Northern Europe: (i) seeds with a smooth chalaza (type *V. teutonica*) and (ii) seeds with a wrinkled chalaza (type *V. ludwigii*). The *teutonica* type are somewhat similar to modern *Vitis* but seeds of *ludwigii* resemble those of *Muscadinia* or *Ampelocissus*. The occurrence of these fossils suggests that the genus *Vitis* was widely distributed in the Northern Hemisphere by the end of the Tertiary period.

The breaking of the intercontinental bridge during the Quaternary period separated the area of primitive *Vitis* into two parts, the American and the Eurasian. Some American and Eurasian species exhibit striking similarities in morphology. *Vitis labrusca* from America has a strong resemblance to the Asian species *Vitis coignetiae*. Similarly, *Vitis caribeae* and *Vitis lanata* have apparent affinities (Levadoux *et al.*, 1962). It is tempting to conclude that these pairs of species arose from a common ancestor and then evolved in equivalent niches in each of the two continents. However, it is equally possible that resemblances between American and Eurasian species are due to similar adaptations by different progenitors. In this regard, the affinities between *V. bourgeana* from Mexico and *V. reticulata* from western China may be adaptations to desert environments rather than indications of common origin.

Most species of *Vitis* are thought to have arisen during the Quaternary ice ages. According to De Lattin (1939) the distribution of contemporary *Vitis* is consistent with the breaking up of large populations by the ice-fronts and with the survival of small populations in *refuges*, or areas which were protected from glaciation by topography or geology. The isolation and differing environmental conditions of the refuges provided ideal circumstances for speciation. During the interglacial periods it is likely that there was expansion and coalescence of the refuges, and re-selection for adaptation to the new warmer environment. The formation and dissolution of *Vitis* refuges probably occurred several times during the Quaternary period.

RESISTANCE TO PESTS AND DISEASES

The two continents formed during the Quaternary period each developed many distinct environments, and a distinct flora and fauna evolved within each environment. Included were the *Vitis* species and their pests and pathogens (fungi, bacteria, viruses). North American species are mostly resistant or tolerant to phylloxera (*Daktulosphaira vitifoliae* FITCH) but *V. vinifera* is highly susceptible. There is some variation among American species (Boubals, 1966). Phylloxera occurs throughout the United States, including the south-western states of

New Mexico, Arizona and California. The level of tolerance in *Vitis berlandieri*, a species that arose in an environment that favors phylloxera (the calcareous soils of central Texas) is much greater than in species such as *V. labrusca* and *V. aestivalis*, which arose in sandy habitats in the Appalachian mountains, environments that are unfavorable to the root-living form of phylloxera. Another example of the adaptation of host to pathogen is the high level of tolerance to Pierce's Disease in *Vitis* species native to the south-east of the U.S.A. (*V. coriacea* and *V. simpsonii*) where the causal organism is endemic (Mortensen *et al.*, 1977).

BARRIERS TO INTERBREEDING

As a result of evolutionary pressures during the Quaternary ice ages, the species of *Vitis* acquired a remarkable diversity in morphological and physiological characters, but this was not associated with significant genetic differentiation. There are no genetic barriers in the genus *Vitis* and the species are interfertile. Numerous interspecific hybrids have been produced by plant breeders and these hybrids are invariably fertile and vigorous. There have been no reports of sexual sterility or break-down in the first or following generations. This is surprising because the geographical isolation of many *Vitis* species, particularly those of American and Eurasian origin, is very ancient. With regard to natural hybridization, wild grapevines are mostly dioecious, so ensuring cross-pollination by wind or insects, and hybridity in *Vitis* is favored by the overlapping distribution of species. In parts of central Texas as many as eight species overlap in their distribution, but there are some natural barriers to interbreeding. For example, there are ecological barriers between sympatric species when their habitats are markedly different. *V. cordifolia* and *V. monticola* are both found in Texas. The former grows in wet bottom-lands and is highly susceptible to lime-induced chlorosis, but the latter grows only on dry, sunny, calcareous hillsides. There are also phenological barriers to interbreeding. Among American species *V. riparia* is early-flowering, *V. cordifolia* is intermediate in its time of flowering and *V. cinerea* is late-flowering. In south-east Kansas these three species grow in the same habitat (woods and thickets adjacent to water-courses), but they do not produce natural hybrids (Levadoux *et al.*, 1962).

THE DEFINITION OF SPECIES

As will be clear from the foregoing, the concept of a species in the genus *Vitis* is less well-defined than in most other crop plants. Species of *Vitis* are ecospecies and can be defined as populations of grapevines that are easily distinguishable by morphological characters and which are isolated, the one from the other, by geographical, ecological or phenological barriers. Thus, each species represents the outcome of adaptation to specific environmental conditions.

The genus *Muscadinia* (PLANCH.) SMALL

INTRODUCTION

This genus, formerly *Vitis* sect. *Muscadinia* PLANCH., contains only three species. *M. rotundifolia* (MICHX.) SMALL, the muscadine grape, is a native of the south eastern United States and it is cultivated as a fruit-crop. A few thousand hectares of muscadine grapes are grown in the area of the Cotton Belt. *M. munsoniana* (SIMPSON) SMALL is found along the coast of the Gulf of Mexico from Texas to Florida but it is not a cultivated species. *M. popenoei* FENNEL is a little-known species, which is found in Mexico.

TAXONOMY OF *MUSCADINIA*

The genera *Vitis* and *Muscadinia* are easily distinguished on the basis of morphological, anatomical and karyological characters (Table 2.3) (Small, 1903). There are no naturally occurring hybrids, and artificial hybridization between *V. vinifera* and *M. rotundifolia* is difficult. The genetic barrier between the two genera is related to differences in chro-

Table 2.3. *Differences between the genera* Vitis *and* Muscadinia

	Vitis	*Muscadinia*
Lenticels	absent	present
Tendrils	forked	simple
Seeds	ovoid-shaped	oblong-shaped
	smooth chalaza	wrinkled chalaza
Pith	discontinuous	continuous
Phellogen	deep-seated	subepidermal
Phloem fibers	tangential	radial
Specific gravity of the wood	< 1	> 1
Chromosome number	$2n = 38$	$2n = 40$
Number of species identified	59	3

mosome numbers and lack of chromosome homology. Cytological analysis of F_1 hybrids of *V. vinifera* and *M. rotundifolia* by Patel and Olmo (1955) showed that the two genera have only 13 pairs of chromosomes in common and they suggested the following genomic structures:

V. vinifera	$13\,R^vR^v + 6AA$
M. rotundifolia	$13\,R^rR^r + 7BB$
F_1-hybrids	$13\,R^vR^r + 6A + 7B$

In this model the basic chromosome numbers in the family are probably 6 and 7. *Vitis* species are thus ancient secondary polyploids involving three basic sets in the combination $(6 + 7) + 6 = 19$. *Muscadinia* species are $(6 + 7) + 7 = 20$. Both genera have undergone diploidization (*Vitis*: $2n = 38$ and *Muscadinia*: $2n = 40$) to give regular pairing. Analyses of F_1 hybrids of *V. vinifera* and *M. rotundifolia* have also been made by Jelenkovic and Olmo (1968) and Bouquet (1980) and the results of these studies support the polyploid origin of muscadine and *vinifera* grapes.

ORIGIN OF *MUSCADINIA*

Muscadinia occur only in North America, but *V. ludwigii*-type fossil seeds are found in the Tertiary sediments of Northern Europe and these seeds bear a strong resemblance to modern *Muscadinia*. This suggests that the genus may have been distributed throughout the Northern Hemisphere before the advent of the ice ages. Accordingly, the separation of *Vitis* and *Muscadinia* may date from the Tertiary period and *Muscadinia* could be regarded as transitional between *Vitis*, which are adapted to temperate climates, and *Ampelocissus*, which are adapted to tropical climates. Species of each of the three genera *Vitis*, *Muscadinia* and *Ampelocissus* have marked similarities in morphology, anatomy and karyotype but little is known of their ability to hybridize. It is probable that *Muscadinia* became extinct in Europe during the Quaternary. So far, there is no evidence that *Muscadinia* occurred in east Asia.

MUSCADINIA GERMPLASM IN GRAPEVINE BREEDING

The discovery of hermaphrodite mutants by Dearing (1917) led to the breeding of numerous perfect-flowered, high-yielding cultivars, but the growing of muscadine grapes has not spread beyond the southern states of the USA. In terms of world viticulture *Muscadinia* are of greatest interest as germplasm in the breeding of hybrid rootstocks and scion cultivars with resistance to pests and diseases. The production of hybrids between *Vitis* and *Muscadinia* is difficult. For reasons that are unknown, crosses using *M. rotundifolia* as the female parent are usually unsuccessful. F_1 hybrids can be obtained by using *M. rotundifolia* as the male parent, but success varies with the genotype of the *V. vinifera* female. The F_1 hybrids are mostly sterile owing to inadequate pairing at meiosis, but fertile F_1 amphidiploids can be produced by colchicine treatment (Dermen, 1964). Partially fertile F_1 hybrids were isolated by Jelenkovic and Olmo (1968) and these were successfully back-crossed to *Vitis vinifera*, thus making it possible to exploit *Muscadinia* germplasm in grapevine improvement (Olmo, 1971).

Cultivars of grapevines

WINE, TABLE GRAPE AND RAISIN CULTIVARS

The production of wine, table grapes and raisins is based primarily on traditional cultivars of *Vitis vinifera*, which have been perpetuated for centuries by vegetative propagation. There has been some progress in the breeding of new rootstocks, but plant breeding has made little impact on viticulture at the level of the scion. The reasons for the persistence of the traditional cultivars of grapevines are many and involve a complex mixture of plant and human factors. This subject will be discussed in detail in Chapter 7.

There are approximately 10 000 known cultivars of *vinifera* grapes. Included are genotypes which are widely grown in many grape growing countries and those which are of strictly local importance.

A few cultivars of wine grapes have achieved great international prominence, for example, Cabernet Sauvignon, Chardonnay, and Pinot noir among those of French origin, and Riesling from Germany (Table 2.4). Cabernet Sauvignon, the famous red wine grape from Bordeaux, is now widely grown in western and eastern Europe, California, Australia, South Africa, Chile and Argentina. The cultivar Thompson Seedless (also known as Sultanina and Kis-Mis), is of major importance as a table grape, and for raisin production, in Europe, America, the Middle East, North and South Africa and Australia.

By contrast, some cultivars of grapevines are very limited in their distribution. In most European wine growing countries individual regions have their own distinctive, traditional cultivars. An example is Piedmont in northern Italy (Table 2.5). Generally, the cultivars of Piedmont are not much grown in other parts of Italy or in other coun-

Table 2.4. Some well-known cultivars of vinifera grapes

Cultivars for red wines		Cultivars for white wines		Cultivars of table grapes and raisin grapes	
name	origin	name	origin	name	origin
Aramon	S. France	Chardonnay	Burgundy, Champagne, France	Almeria	Spain
Cabernet franc	Bordeaux, France			Black Corinth (syn. Zante currant)	Ancient, Greece
Cabernet sauvignon	Bordeaux, France	Chenin blanc	Loire Valley, France		
Carignane	S. France	Colombard	Charente, France	Cardinal	Fresno, California (1938)
Gamay	Beaujolais, France	Traminer	Germany		
Grenache	Rhône Valley, France	Riesling	Germany	Chasselas doré	S.W. France
Merlot	Bordeaux, France	Sauvignon blanc	Loire Valley, France	Waltham Cross (syn. Dattier de Beyrouth)	Middle East
Mataro	Spain, S. France	Semillon	Bordeaux, France		
Nebbiolo	N. Italy	Sylvaner	Germany, Austria		
Pinot noir	Burgundy, Champagne, France	Müller-Thurgau	Germany	Emperor	Ancient, Middle East
		Emerald Riesling	Davis, California	Flame Tokay	Ancient, Spain/ N. Africa
Syrah (syn. Shiraz)	Rhône Valley, France	Ugni-Blanc	Italy	Italia	Italy (1911)
Ruby Cabernet	Davis, California			*Muscat of Alexandria	Ancient, Middle East
Carmine	Davis, California			Muscat Hamburg	France
				Perlette	Davis, California (1936)
				Ribier	Belgium (1860)
				*Thompson Seedless (syn. Sultanina)	Ancient, Middle East

* Also used to produce raisins.

Table 2.5. *Traditional wine grape cultivars of Piedmont, Italy*

Cultivar	Locality where grown
Barbera (red)	Asti, Allesandria, Cuneo, Novara
Nebbiolo (red)	Widespread in Piedmont
Dolcetto (red)	Mondovi, Alba, Asti, Ovada, Nova Ligure
Grignolino (red)	Asti, Casale
Friesa (red)	Asta, Allesandria, Turin, Cuneo, Chieri
Cortese (white)	Allesandria, Acqui, Tortona, Gavi, Nova Ligure
Erbaluce (white)	Torino, Caluso

Tosatti and Minetti (1988).

tries. Within Piedmont, the growing of some traditional cultivars is further restricted to particular localities for reasons of history or of special environmental adaptation.

ROOTSTOCK CULTIVARS

Viticulture is based primarily on the use of grafted plants in which the scion is a cultivar of *Vitis vinifera*, and the rootstock is either a North American species or interspecific hybrid, which is resistant to the soil-borne pests such as phylloxera or nematodes. There are a few exceptions to these generalizations. In parts of North America with very cold winters, scion cultivars are usually interspecific hybrids between *V. vinifera* and cold-hardy native species.

The majority of Australian viticulture is in phylloxera-free regions and grapevines are grown on their own roots. The history, properties and breeding of grapevine rootstocks will be discussed at length in other chapters. Meanwhile, the principal rootstock cultivars are listed in Table 2.6.

THE EUROPEAN GRAPE, *VITIS VINIFERA L.*

THE ORIGIN OF TRADITIONAL CULTIVARS

The center of origin of *Vitis vinifera*, in the sense of Vavilov's 'centers of diversity', was the Transcaucasian region between the Black Sea and the Caspian Sea, and it is in this region that wild grapevines are still the most abundant and the most variable (Negrul, 1936). However, there is archaeological evidence that wild grapevines were distributed throughout Europe by the end of the Pleistocene and that grapes were consumed by Neolithic people. As discussed earlier, the technology of viticulture

Table 2.6. *Grapevine rootstocks: some long-established and newly bred cultivars*

Rupestris St George (du Lot)	*V. rupestris*
Riparia Gloire de Montpellier	*V. riparia*
3309 Courderc	*V. riparia* × *V. rupestris*
101-14 Millardet et de Grasset	*V. riparia* × *V. rupestris*
Schwarzmann	*V. riparia* × *V. rupestris*
99 Richter/110 Richter	*V. berlandieri* × *V. rupestris*
140 Ruggeri	*V. berlandieri* × *V. rupestris*
1103 Paulsen	*V. berlandieri* × *V. rupestris*
5BB Teleki, selection Kober	*V. berlandieri* × *V. riparia*
SO4	*V. berlandieri* × *V. riparia*
420A Millardet et de Grasset	*V. berlandieri* × *V. riparia*
Ramsey	*V. candicans* × *V. rupestris*
Börner	*V. cinerea* × *V. riparia*
Vialla	*V. labrusca* × *V. riparia*
Ganzin 1 (AxR # 1)	*V. vinifera* × *V. rupestris*
41B Millardet et de Grasset	*V. vinifera* × *V. berlandieri*
Fercal	*V. vinifera* × *V. berlandieri*
039-16	*V. vinifera* × *Muscadinia rotundifolia*
043-43	*V. vinifera* × *Muscadinia rotundifolia*

Pongracz (1983); Galet (1979).

and winemaking probably arose in Transcaucasia and then spread westward, and this may have included the dissemination of cultivated varieties of grapevines. Nevertheless, some of the cultivars now grown in Europe may have originated from native grapevines rather than from introduced grapevines. Wild grapevines were common in Europe until the nineteenth century, but they probably became extinct as a consequence of infestation by phylloxera.

Gmelin (1805, cited by Levadoux, 1956) classified the wild grapevines of the Rhine valley as a separate species, *Vitis sylvestris*. The two main characters used to differentiate the wild *sylvestris* from the cultivated *vinifera* were dioecism and seed morphology: *sylvestris* are dioecious (male and female plants) but *vinifera* are hermaphrodite; the seeds of *sylvestris* are rounded in outline, but the seeds of *vinifera* are elongated in outline, particularly in the region of the beak. The status of *Vitis sylvestris* C. C. GMELIN, also known as *Vitis vinifera* L. subspecies *sylvestris* (C. C. GMELIN) HEGI, is controversial; some authorities do not accept that *V. sylvestris* is different from *V. vinifera*.

THE CLASSIFICATION OF CULTIVARS

Levadoux (1956) classified the main cultivars of grapevines grown in France according to their morphological similarities and geographical distribution. He identified several groups or families of cultivars and postulated that resemblances among cultivars within a family were an indication that the genotypes concerned had arisen either from the same population of wild grapevines or from the same population of introduced grapevines. In the classification of Levadoux the family *Noiriens* contains well-known cultivars such as the Pinots, Chardonnay, Meunier, the Gamays, Melon, Sauvagnin and Traminer and lesser known cultivars such as Teinturier du Cher.[3]

A criticism of the classification of Levadoux is that it is difficult to establish the true origin of many cultivars of grapevines grown in western Europe because the wild progenitors are extinct. There are a few small populations of so-called wild grapevines in the Rhine Valley and the French Pays Basque. Studies on the distribution of flower types in the wild grapes of the Pays Basque led Carbonneau (1983) to the conclusion that many wild-growing grapevines were feral, that is, they were derived from plants that were once cultivated rather than from wild plants in the true sense.

THE CLASSIFICATION OF NEGRUL: THE GROUPING OF CULTIVARS INTO 'PROLES'

Negrul (1946) distinguished three main groups of grapevine cultivars or proles: *Proles pontica, Proles orientalis* and *Proles occidentalis.*

Proles pontica comprises the oldest of cultivars, which arose close to the center of origin of viticulture in Transcaucasia and which then spread to the Balkans, the 'Fertile Crescent' and Egypt. Two sub-proles can be distinguished: (i) *georgica*, included in which are the cultivars Rkatziteli, Saperavi, Odjalechi and Mtsvane; and (ii) *balkanica*, included in which are the cultivars Furmint, Clairette, Harslevelu, Vermentino and Zante currant (Black Corinth).

Grapevines that were introduced into the Middle East during the period 3000–2000 BC were probably *Proles pontica*. Cultivars from the region of the Caspian Sea are so different from *Proles pontica* that Negrul proposed that they arose from a different form of the so-called *Vitis sylvestris*, i.e. *V. sylvestris aberrans* as distinct from *V. sylvestris typica*. These cultivars, which include the grapevines used for winemaking before the advent of Islam, were attributed to the *Proles orientalis sub-proles caspica*.

---

[3] The term 'teinturier' refers to grapevines in which the pulp cells of the fruit contain anthocyanin pigments. These genotypes produce an intensely colored red juice. The cultivar Teinturier du Cher was selected from a population of wild grapevines in northern France and it was used by breeders during the period 1820–50 to produce high yielding teinturiers for the south of France. The best known of these red-juiced cultivars is Alicante Bouschet.

Modern cultivars that belong to this group include the Muscats and Cinsaut. The *Proles orientalis sub-proles antasiatica* are mostly table grape cultivars. These genotypes arose by selection from wine grapes belonging to *sub-proles caspica* during the time of Islamic influence in the Mediterranean basin (AD 500–1100). The best-known and most widely grown cultivar of the *sub-proles antasiatica* is Sultanina (syn. Thompson Seedless).

The earliest cultivars in western Europe were introduced by the Phoenicians and Greeks and the grapevines concerned probably belonged to *Proles pontica* or to *Proles orientalis sub-proles caspica*. These cultivars were unsuitable for the northerly areas of Europe, and cultivars with better adaptations to cool conditions were selected from populations of native wild grapevines. These cultivars comprise the *Proles occidentalis* within which are most of the better-known wine grapes of France and Germany.

RELATIONSHIP BETWEEN MODERN GRAPEVINES AND CULTIVARS OF THE ROMAN ERA AND MIDDLE AGES

Roman historians referred to many cultivars of grapevines. Included were the *Allobrogica* of the Rhône valley and the *Biturica* of Bordeaux. Some historians have tried to link these Roman cultivars to present-day grapevines. The *Allobrogica*, for example, is said to be the ancestor of the Pinots. In the patois of the Bordelais the name for Cabernet Sauvignon is '*Bidure*' or '*Vidure*,' and these names are said to be corruptions of *Biturica*, a grapevine described by Pliny. In fact, most of these linkages are flights of fancy because the descriptions of grapevines by Roman authors are inadequate for the purposes of cultivar identification. A possible exception is the *Apianae* vines of Rome, which are likely to have been muscats (Bouquet, 1982).

With the fall of the Roman Empire, cultivated grapevines reverted to the wild and interbred with native grapevines. Therefore, it is likely that the wild grapevines found throughout Europe from the Middle Ages until the nineteenth century contained germplasm from Roman or Gallo-Roman cultivars and from native *Vitis vinifera*. The speed with which exotic germplasm can escape from cultivation is illustrated by the occurrence at the end of the nineteenth century of wild populations of *Vitis riparia* and *Vitis rupestris* along the Rhône and Garonne rivers. These North American species were first introduced into France in the 1860s for use as phylloxera-resistant rootstocks, and their seed was spread by birds.

AMPELOGRAPHY: THE IDENTIFICATION AND CLASSIFICATION OF GRAPEVINE CULTIVARS

The term *ampelography* refers to the art and science of grapevine description and identification. The subject of ampelography is French in origin and its practitioners are known as *ampelographers*. Books that describe grapevines are called *ampelographies*. Before the eighteenth century the writing of ampelographies, like the writing of herbals, was associated with the monasteries. Wine is of great significance in the Christian Mass, and the culture and study of grapevines were important aspects of religious life. Later, ampelography became an enthusiasm of the aristocracy, but it was not until the turn of the twentieth century that the identification of grapevine cultivars attracted the attention of botanists. One of the best known classical ampelographies is that of Viala and Vermorel (7 volumes), published in France during the period 1900–10.

Viticulture is based on clones and on the perpetuation of specific genotypes by vegetative propagation. The antiquity of grape growing, and the fact that rootings or hardwood cuttings can readily be transported over long distances, has led to much confusion in the naming of cultivars. The same cultivar may have many different names and different cultivars may have the same name. For example, the cultivar known as Ugni-Blanc in the south of France originates from Italy, where it is called Trebbiano Toscano. In the Cognac area there are 85 000 ha of Ugni-Blanc, the principal cultivar for cognac production, but it is known locally as St Emilion des Charentes. Another example of confusing nomenclature is the table grape cultivar Waltham Cross, which is known as Regina in Italy, Dattier in France, Bolgar or Afuz-Ali in Bulgaria, Rozaki in Greece and Turkey, and Real or Teta de Vaca in Spain.

The name Riesling has been assigned to many different cultivars. The authentic Riesling is the famous white-wine cultivar grown in Germany and Alsace. In Austria, Riesling is called Rhein Riesling to distinguish it from Welch Riesling, an inferior cultivar. The Riesling cultivated in South Africa and the Clare Riesling grown in Australia are, in fact, the cultivar Crouchen, which comes from the south west of France. In Australia, the cultivar known as Hunter River Riesling is Semillon, and in California the cultivar Gray Riesling is the same as the French cultivar Trousseau.

In modern times, ampelography has been refined to become an objective method of cultivar and species identification and it is based primarily on *ampelometry*, measurements on grapevine leaves. The development of modern ampelography was prompted in part by the highly confused nomenclature of the 10 000 or more known cultivars of *Vitis vinifera* and in part by the introduction of new pests and diseases into Europe from North America in the latter part of the nineteenth century. During this time much germplasm was introduced into France from the USA, and great effort was given to producing interspecific hybrids, both rootstocks and scions, with resistance to phylloxera, downy and powdery mildew and black rot. The need arose for accurate descriptions of introduced materials and for a means of positive identification of newly produced hybrids. In particular, it was necessary to distinguish newly bred cultivars from the traditional cultivars of wine grapes, the growing of which was, and still is, subject to law in France and other European countries.

The best known and most straightforward ampelographic procedure is that of Galet (1971, 1979) in which cultivars are identified on the basis of two main morphological characteristics, the shape of the leaf and the pubescence of the shoot-tip. Leaf shape is expressed in terms of the ratios of the lengths of lateral veins to the length of the main vein and by the sum of the angles formed by certain veins. The description of each cultivar comprises a drawing of the outline of a 'typical' fully expanded leaf and a brief statement containing quantitative and qualitative information. By use of ampelographic procedures Truel (1985) was able to show that 1400 introduced cultivars in the germplasm collection of INRA (Institut National de la Recherche Agronomique, France) were misnamed and were synonymous with existing cultivars.

A complex ampelographic procedure was proposed by the Office International de la Vigne et du Vin in 1951, which was based on 65 morphological characters and 267 levels of expression, but methods such as this were little used until the advent of computers and electronic data storage and retrieval (Boursiquot *et al.*, 1987). Other numerical taxonomic approaches to the identification of grapevine cultivars involving multivariate analysis have been developed (Fanizza, 1980). In the future, ampelographic methods are likely to be based on molecular differences among cultivars rather than on morphological characters, which are often affected by environment or plant age.

Literature cited

Bailey, L.H. 1934. The species of grapes peculiar to North America. *Gentes Herbarum* **3**: 151–244.
Boubals, D. 1966. Etude de la distribution et des causes de la résistance au phylloxera radicicole chez les Vitacées. *Ann. Amélior. Plantes* **16**: 145–84.

Bouquet, A. 1980. *Vitis* × *Muscadinia* hybridization: A new way in grape breeding for disease resistance in France. *Proc. Third Int. Symp. Grape Breeding, Univ. Calif., Davis*, pp. 42–61. University of California, Davis.

Bouquet, A. 1982. Origine et évolution de l'encépagement français à travers les siècles. *Prog. Agric. Vitic.* **99**: 110–21.

Boursiquot, J.M., Faber, M.P., Blachier, O. and Truel, P. 1987. Utilisation par l'informatique et traitement statistique d'un fichier ampelograhique. *Agronomie* **7**: 13–20.

Cahoon, C.A. 1986. The Concord grapes. *Fruit Var. J.* **40**: 106–7.

Carbonneau, A. 1983. Stérilités male et femelle dans le genre. *Vitis*: I. Modélisation de leur hérédité. *Agronomie* **3**: 635–44.

Chadefaud, M. and Emberger, L. 1960. *Les Végétaux Vasculaires. Traité de Botanique*, vol. 2. Masson, Paris. 1540 pp.

De Lattin, G. 1939. Über den Ursprung und die Verbeitung der Reben. *Züchter* **11**: 217–25.

Dearing, C. 1917. Muscadine grape breeding. *J. Heredity* **3**: 409–24.

Dermen, M. 1964. Cytogenetics in hybridization of bunch and muscadine-type grapes. *Econ. Bot.* **18**: 137–48.

Fanizza, G. 1980. A numerical taxonomic approach to the ampelography of *vinifera* wine grapes. *Proc. Third Int. Symp. Grape Breeding, Univ. Calif. Davis*, pp. 99–104. University of California, Davis.

Foex, G. 1895. *Cours complet de viticulture.* Masson, Paris. 1121 pp.

Galet, P. 1967. *Recherches sur les méthodes d'identification et de classification des Vitacées des zones tempérées.* Thèse, Université de Montpellier. 600 pp.

Galet, P. 1971. *Précis d'ampélographie pratique.* Third édition. P. Galet, Montpellier. 266 pp.

Galet, P. 1979. (Translated by L.T. Morton). *Practical ampelography. Grapevine identification.* Comstock Publishing Association, Cornell University Press, Ithaca, New York. 248 pp.

Huang, H.B. 1980. Viticulture in China. *HortScience* **15**: 461–6.

Jelenkovic, G. and Olmo, H.P. 1968. Cytogenetics of *Vitis*. III. Partially fertile F_1 diploid hybrids between *V. vinifera* L. and *V. rotundifolia* Michx. *Vitis* **7**: 8–18.

Kirchmeimer, F. 1939. Rhamnales. I. Vitaceae. In *Fossilium catalogus*, volume 2 (*Plantae*), pp. 2–153 Jongmans Faller, Neubrandenburg.

Kliewer, W.M., Nassar, A.R. and Olmo, H.P. 1966. A general survey of the free amino acids in the genus *Vitis. Am. J. Enol. Vitic.* **17**: 112–17.

Lavie, P. 1970. *Contribution à l'étude caryosystématique des Vitacées.* Thèse, Université de Montpellier I, Faculté des Sciences. 292 pp.

Levadoux, L. 1956. Les populations sauvages et cultivées de *Vitis vinifera. Ann. Amélior Plantes* **6**: 59–118.

Levadoux, L. 1968. Essai de regroupement phylogénétique des vignes vraies d'Amérique. *Rev. Hortic. Vitic. (Bucarest)* **7**: 31–28.

Levadoux, L., Boubals, D. and Rives, M. 1962. Le genre *Vitis* et ses espèces. *Ann. Amélior. Plantes* **12**: 19–44.

Mortensen, J.A., Stover, L.H. and Balerdi, C.F. 1977. Sources of resistance to Pierce's Disease in *Vitis. J. Amer. Soc. Hort. Sci.* **102**: 695–7.

Munson, T.V. 1909. *Foundations of American grape culture.* T.V. Munson and Son, Denison, Texas. 252 pp.

Negrul, A.M. 1936. Variabilitat und Vererbung des Geschlechts bei der Rebe. *Gartenbauwiss.* **10**: 215–231.

Negrul, A.M. 1946. Origin of the cultivated grapevine and its classification. (In Russian.) *Ampelographia SSSR, Moscow* **1**: 159–216.

Olmo, H.P. 1971. *Vinifera* × *rotundifolia* hybrids as wine grapes. *Am. J. Enol. Vitic.* **22**: 87–91.

Olmo, H.P. 1978. Genetic problems and general methodology of breeding. In *Génétique et amélioration de la Vigne*, pp. 3–10 INRA, Paris.

Olmo, H.P. 1980. Natural hybridization of indigenous *Vitis californica* and *V. girdiana* with cultivated *vinifera* in California. *Proc. Third Int. Symp. Grape Breeding, Univ. Calif. Davis*, pp. 31–41. University of California, Davis.

Patel, G.I. and Olmo, H.P. 1955. Cytogenetics of *Vitis*. I. The hybrid *V. vinifera* × *V. rotundifolia*. *Am. J. Bot.* **42**: 141–59.

Planchon, J.E. 1887. Monographie des Ampélideae vraies. *Monographia Phanerogamerum* **5**: 305–64.

Pongracz, D.P. 1983. *Rootstocks for grapevines*. David Philip Publisher, Cape Town. 150 pp.

Ravaz, L. 1902. *Les vignes américaines. Porte-greffes et producteurs-directs: caractéres, aptitudes*. Coulet et Fils, Montpellier/Masson et Cie, Paris. 376 pp.

Ribéreau-Gayon, P. 1959. *Recherches sur les anthocyanes des végétaux: Applications au genre Vitis*. Thése, Université de Paris (Libraire générale de l'enseignement, Paris). 114 pp.

Rogers, D.J. and Rogers, C.F. 1978. Systematics of North American grape species. *Am. J. Enol. Vitic.* **29**: 73–8.

Samaan, L.G. and Wallace, D.H. 1981. Taxonomic affinities of five cultivars of *Vitis vinifera* L. as aided by serological analysis of pollen proteins. *J. Amer. Soc. Hort. Sci.* **106**: 804–9.

Schaefer, H. 1971. Endopolymorphismus in Rebenblattern. *Phytochem.* **10**: 2601–7.

Shetty, B.V. 1959. Cytotaxonomical studies in *Vitaceae*. *Bibliogr. Genet.* **18**: 169–264.

Small, J.K. 1903. *Vitis* and *Muscadinia*. In *Flora of the southeastern United States*. J.K. Small, New York. pp. 752–7.

Sussenguth, K. 1953. *Vitaceae*. In *Die Naturlichen Pflanzenfamilien*. **20**: 174–398. Enger and Prantl, Berlin.

Tosatti, R. and Minetti, G. 1988. *Il vino è Piemonte*. Barisone Editore, Torino. 191 pp.

Truel, P. 1985. *Catalogue des variétés de vigne en collection*. Institut National de la Récherche Agronomique, Montpellier. 129 pp.

Tukey, H.B. 1966. The story of the Concord grape. *Fruit Var. Hort. Digest* **20**: 54–5.

Viala, P. and Verrnorel, V. 1910. *Ampélographie*. (7 vols.) Masson et Cie, Paris.

Wolfe, W.H. 1976. Identification of grape varieties by isozyme banding patterns. *Am. J. Enol. Vitic.* **27**: 68–73.

3
The structure of the grapevine: vegetative and reproductive anatomy

Introduction

In the wild, *Vitis vinifera* L. is a vigorous climbing plant of deciduous forest. Its trunk and branches are flexible, and the plant is supported by the trees on which it grows. The climbing habit of the grapevine is reflected in the occurrence of pressure-sensitive tendrils; wild vines climb into the forest canopy to a height of 20–30 m. In the Middle East, cultivated grapevines are often grown along the ground, but viticulture in the European tradition is based either on small free-standing bushes or on the training of grapevines onto supports. In the days of the Romans these supports were living poplar trees; examples of this training system can still be seen in northern Italy. In mechanized viticulture, grapevines are usually trained onto post-and-wire trellises of many different designs. In its wild state the grapevine produces large numbers of small bunches of fruit. As a crop plant the grapevine is severely pruned so as to reduce bunch number and to increase fruit size and fruit quality. The grapevine has a remarkable ability to regrow after pruning and to produce new crops of extension shoots, and this enables the annual renewal of the fruiting wood. Carefully tended grapevines can remain productive for a very long time, if not indefinitely; a familiar example of a long-lived plant is the Great Vine at Hampton Court Palace, London, which was planted in 1769. In commercial viticulture, grapevines are seldom retained for more than 40 years.

Important biological characteristics of *Vitis vinifera* and many of its relatives include vigor, floriferousness, regenerative capacity, stress tolerance and longevity. This chapter is concerned with the structural bases of these characteristics. The vine is a complex plant and its peculiarities have been the subject of scholarly work for more than a century. Only the salient features of structure will be considered here.

37

The shoot system

JUVENILE AND ADULT MORPHOLOGY

The complexities of the grapevine shoot and its appendages can most easily be explained if the point of departure is the germination of the seed rather than bud burst in a vegetatively propagated grapevine. The incipient apical meristem (gemmule) is located between the two cotyledons of the mature embryo and consists of a pocket of approximately 200 meristematic cells (Figs 3.1 and 3.2). During germination these

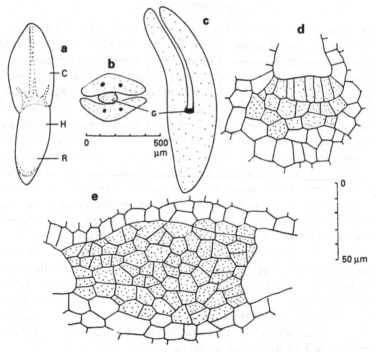

Fig. 3.1. The zygotic embryo. (*a*) Mature embryo. C, cotyledon; H, hypocotyl; R, radicle. The stippled areas represent the vascular elements and the root apical meristem as seen in a cleared specimen. (*b*) Transverse section through embryo at base of cotyledons. (*c*) Longitudinal section through embryo at base of cotyledons. Note the position of the gemmule (G), the incipient apical meristem. (*d*) Longitudinal section through gemmule. (*e*) Transverse section through gemmule. Note the typical meristematic cells. There is no indication of primordium formation at this stage. From Bugnon and Bessis (1968). Reproduced with permission

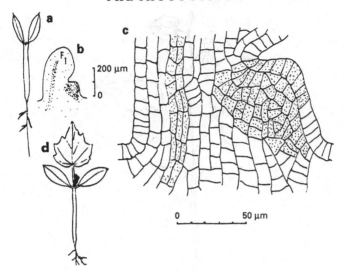

Fig. 3.2. Germination and seedling growth in the grapevine. (a) Early stage of seedling growth. The cotyledons have expanded and are borne on an elongated hypocotyl. (b) Longitudinal section of the gemmule (at same stage as (a)). Note the first leaf primordium (F1) and the apical meristem. (c) Enlargement of (b) to illustrate structure of the apical meristem. (d) Young seedling. The first leaf has expanded. From Bugnon and Bessis (1968). Reproduced with permission

cells enter into active division and give rise to a series of leaf primordia. When the seedling shoot elongates these first-formed leaves exhibit 2/5 phyllotaxy, that is, they are arranged about the axis on a spiral in which the angle between successive leaves is approximately 145°. Within the axils of the first-formed leaves are axillary buds which, in turn, contain two basal scales or prophylls and a series of primordial leaves. There are no tendrils. This is the juvenile morphology of the grapevine (Fig. 3.3).

The juvenile phase is short-lived and there is an abrupt change to the adult morphology after the production of six to ten leaves by the apical meristem. At the transition the phyllotaxy becomes distichous (two-ranked) rather than spiral. In other words, leaves are produced on two opposite sides of the stem so that the shoot is bilaterally symmetrical with respect to leaf production rather than radially symmetrical as in the juvenile. There is also a change in the structure of axillary buds in that there is generally only one prophyll proximal to the first foliage leaf.

Of greatest significance is the appearance of tendrils at positions on the stem opposite to leaves. The production of tendrils denotes the acquisition of the climbing habit by the grapevine, although not neces-

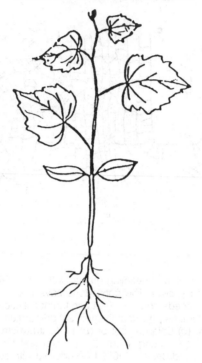

Fig. 3.3. The grapevine seedling after the production of four leaves.
The phyllotaxy is spiral (2/5) and tendrils are absent

sarily the acquisition of sexual maturity. A unique feature of cultivated grapes and their relatives is that tendril formation is discontinuous. There is a repeating pattern of tendril production in which every third node lacks a tendril. Starting at the first tendril-bearing node, the pattern is thus 'tendril, tendril, no-tendril, tendril, tendril, no-tendril' and so forth. In successive cycles the 'missing' tendril occurs on opposite sides of the stem (Fig. 3.4). Other rhythmic or cyclical phenomena within extension shoots of the current season include variations in internode length, diaphragm thickness and lengths of summer lateral shoots. The adult morphology is normally persistent under vegetative propagation but the juvenile morphology may recur when grapevines are grown *in vitro* (Mullins *et al.*, 1979).

Further discussion will be restricted to shoots of the adult phase in which regrowth is from a dormant lateral bud borne on a one-year-old stem (Figs 3.5 and 3.6). The shoot concerned will be referred to as the primary shoot, the shoot of the current season, or the cane.

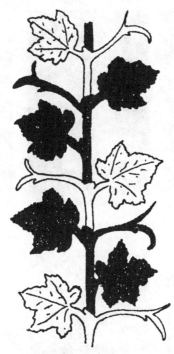

Fig. 3.4. The adult grapevine stem is composed of two morphological
units: (i) a tendril and two leaves and (ii) a tendril and one leaf

THE SHOOT OF THE CURRENT SEASON

The apical meristem of the grapevine comprises an outer tunica of two
layers of cells, which cover a less well defined corpus. The plane of cell
division in the tunica is generally anticlinal (perpendicular to the surface) whereas division in the corpus is both periclinal (parallel to the
surface) and anticlinal. The first leaf primordium arises from initials in
the second layer of the tunica.

The tip of the elongating shoot is usually triangular in outline and is
composed of the apical meristem, leaf and tendril primordia and young
unexpanded leaves and tendrils. Depending on the genotype the shoot
tips may be pigmented with anthocyanin or bear a covering of downy
hair. There are also differences among genotypes in the conformation of
the shoot tip, owing to differences in the rate of unfolding of the newly-
formed leaves. These characteristics of color, hairiness and form are
used as criteria in cultivar identification.

In deciduous fruit trees, such as the apple, extension growth is termi-

Fig. 3.5. Shoot growth of the adult phase is from a dormant lateral bud borne on a one-year-old stem (a). The remaining photographs demonstrate the development of the bud from swelling (b) to bud burst (c) and then shoot growth (d, e, and f). Also see Fig. 4.7 for shoot development sequence. ((a) Bar = 5 mm; (c–e), bar = 10 mm; (f), bar = 20 mm.)

nated by the formation of bud scales. These scales enclose and protect the apical meristem, and regrowth in the next season occurs from well-differentiated terminal buds. In the grapevine there are no 'terminal buds' because the distal portion of the shoot or cane lignifies, but the vascular tissue fails to enlarge and it dies back during winter. Regrowth occurs from the uppermost lateral bud on the lignified ('ripened') portion of the cane. In this sense the vine shoot is a sympodium, but the

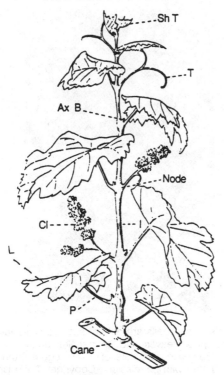

Fig. 3.6. Shoot of *V. vinifera* at bloom, showing the arrangement of leaves, clusters (Cl), and tendrils (T). The axillary buds (Ax B) develop into summer lateral shoots, each bearing a latent bud on the overwintering cane. (L, lamina; I, internode; P, petiole; Sh T, shoot tip). Redrawn from Pearson and Goheen (1988)

primary shoot grows as a single axis with lateral appendages, that is, as a monopodium.

The shoot of the current season is formed by a combination of fixed growth and free growth. Fixed growth refers to the elongation of internodes and the expansion of leaves which were pre-formed in the dormant bud. Free growth refers to the elongation of a shoot by continuous production of new leaf primordia by the apical meristem. Fixed growth accounts for up to 12 of the first-produced nodes of the cane.

The fully elongated internode is elliptical in cross section. The epidermis bears stomata and epicuticular wax, and it is photosynthetic. Under conditions of high temperature and humidity, small globular excrescences known as 'sap balls' or 'pearls' appear on the stem and also

Fig. 3.7. Pearls or sap balls are natural excrescences found on the undersides of grape leaves and on petioles during spring. They are not to be confused with mite or insect eggs. From Kasimatis (1982). Reproduced with permission. (Copyright © the University of California.)

on the petioles, leaves and tendrils. The structure and function of the pearls is not understood but they should not be confused with insect eggs (Fig. 3.7).

The nodes of the vine cane are of greater diameter than the internodes and the extent of the nodal swelling varies with genotype. A diagnostic feature of *Vitis* is the presence of a hard, lignified pith, or diaphragm, at each node, resulting in a discontinuous pith (Fig. 3.8). In *Muscadinia* the diaphragm is absent and the pith is continuous.

PROMPT BUD, LATENT BUD AND SUMMER LATERAL

Historically, the grapevine was not a plant of the English-speaking world and there are no common names in the English language for the parts of grapevines (or, indeed, for many techniques in viticulture and winemaking). In describing the buds of grapevines the use of botanical nomenclature, although technically correct, is very cumbersome. For

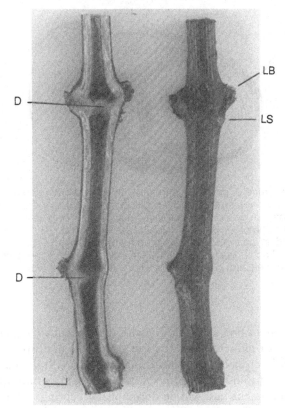

Fig. 3.8. Longitudinal section of a one-year-old vine cane. Note the swollen nodes and the position of the latent bud (LB) directly above the leaf scar (LS). The presence of a diaphragm (D) at each node is a characteristic of *Vitis*. The diaphragm is absent in *Muscadinia*. (Bar = 1 cm.)

convenience, reference will be made here to 'prompt buds' and to 'latent buds,' terms which have been borrowed from the French. The structures concerned are special types of lateral buds, they are specific to grape-vines and their ontogeny is complex (Fig. 3.9).

The first-formed bud that arises in each leaf axil is the prompt bud (prompt bourgeon). This bud grows out in the season of its formation to produce a short lateral shoot known as the summer lateral (entre-coeur, rameau anticipé). The summer lateral is seldom fertile, it usually fails to lignify and it abscises during autumn or winter to leave a prominent scar (cicatrice). The first leaf of the summer lateral is reduced to a prophyll. The bud formed in the axil of this prophyll is the latent bud (bourgeon

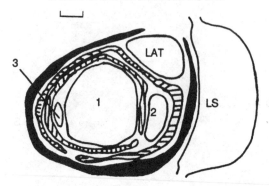

Fig. 3.9. Transverse section through a compound bud (eye) of Concord grape showing the relative positions of the leaf scar (LS), lateral shoot scar (LAT), and three dormant buds (1–3): 1, the primary bud, in the axil of the prophyll (solid black) of the lateral shoot: 2, the secondary bud, in the axil of the basal prophyll (horizontally hatched) of the primary shoot: 3, the tertiary bud, in the axil of the next higher prophyll (vertically hatched) of the primary shoot. Redrawn from Pratt (1974). (Bar = 0.5 mm.)

latent). The latent bud grows slowly within the prophyll; depending on the cultivar, it produces six to ten leaf primordia and up to three inflorescence primordia before entering into dormancy. It should be noted that embryonic prompt buds formed in the axils of leaf primordia within the compound latent bud will lie dormant until spring, when the bud begins growing. The apex of the primary or first-formed latent bud produces two or more bracts before forming leaf primordia. The buds formed in the axils of these bracts are the secondary and tertiary latent buds. These buds remain small and seldom contain inflorescence primordia (Fig. 3.9).

To recapitulate, the primary, secondary and tertiary latent buds are enclosed by the basal bract or prophyll of the summer lateral and by the two basal bracts of the primary latent bud. Together, these structures constitute the prominent latent bud or eye (l'oeil) on a mature vine cane (Figs 3.5 and 3.10). This bud is referred to in the singular as 'the latent bud' but it is a compound bud composed of several buds, each located in the axil of the other. At first sight, the latent bud appears to be axillary to the cane or primary shoot, but it is a basal appendage of the summer lateral and it comprises axillary buds of the second and third orders with reference to the primary shoot. The association between the latent bud and summer lateral is very close; xylem vessels from young latent buds lead directly to the summer lateral.

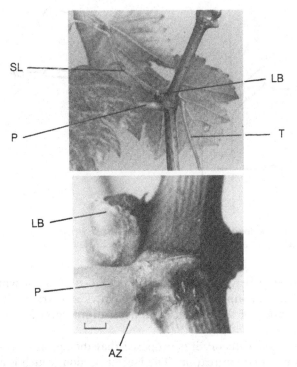

Fig. 3.10. The shoot of the current season has several appendages. Above: vine cane in summer. LB, latent bud; SL, summer lateral; T, tendril; P, petiole. (Bar = 1 cm.) Below: vine cane in autumn. LB, latent bud; P, Petiole. Note abscission zone (AZ). (Bar = 2.5 mm.)

THEORETICAL ASPECTS OF VASCULAR ANATOMY

A tenet of morphology is that there are three components of a higher plant: stem, leaf and root, and that other organs are modifications of these basic components. The interpretation of the tendril-bearing shoots of the grapevine has attracted the attention of comparative morphologists for more than a century, and much effort has been given to establishing homologies between the organs of the primary shoot and between grapevines and other plants (Bugnon and Bessis, 1968).

It is generally agreed that the tendrils of grapevines correspond to determinate leaf-bearing shoots (Tucker and Hoefert, 1968). The shoot axis is the hypoclade (H) which bears two branches, the so-called inner and outer arms, and leaves are represented by a bract (B) (Fig. 3.11).

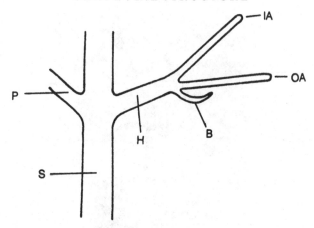

Fig. 3.11. The tendril consists of four structures: hypoclade (H), bract (B), inner arm (IA) and outer arm (OA). S, stem; P, petiole

The relationship between the tendril and the stem upon which it is borne has been the subject of much argument. The 'extra-axillary, leaf-opposed' position of the tendrils is difficult to interpret and there are several theories.

In the sympodial theory it is proposed that the apical meristem undergoes an unequal bifurcation. The largest portion, which is adjacent to the most recently formed leaf primordium, elongates and continues as the main axis. The smaller portion is then displaced laterally by the elongating axis and becomes a tendril primordium. This sequence is repeated at the time of formation of each node and the stem so produced is a sympodium. To account for the fact that tendrils are 'missing' at every third node, it was suggested by Eichler (1878) that the stem is composed of two morphological 'units', the first comprising a terminal tendril and two leaves and the second comprising tendrils and a single leaf (Fig. 3.4).

Evidence for the sympodial arrangement has been derived primarily from studies in comparative morphology; there is no direct anatomical evidence for the lateral displacement of the incipient tendril primordium. Moreover, studies on growth correlations (Antcliff et al., 1957; Bessis, 1965) have shown that only the buds on the same side of the stem are influenced by each other. This suggests that buds on the same side are directly connected to the vascular system. Such a relationship would be unlikely if the stem was constructed as a sympodium. The sympodial theory has had some adherents (Snyder, 1933; Alleweldt, 1963; Alleweldt and Balkema, 1965) but greatest support is now given to the monopodial theory, the main proponent of which has been the French morphologist F. Bugnon (1953; Bugnon and Bessis, 1968).

In the monopodial theory it is hypothesized that the entire annual shoot is derived from the same apical meristem and that leaves, tendrils and inflorescences are lateral appendages of the one axis. The following additional theory has been proposed by Bugnon to account for the extra-axillary, leaf-opposed position of the tendrils: (i) the bract is, in fact, a foliage leaf which has been displaced from the main shoot axis by elongation of the hypoclade; (ii) this foliage leaf is alternate with the foliage leaves of the main axis; and (iii) the internode proximal to the tendril does not expand and this results in the tendril appearing to be paired with the foliage leaf of the proximal node. The monopodial theory is supported by anatomical evidence (Barnard, 1932; Bugnon, 1953; Branas, 1957) and by analysis of chimeras (Breidner, 1953).

A still more complex interpretation is based on Plantefol's (1949) theory of phyllotaxy involving multiple foliar helices. This interpretation alone takes account of the missing tendril at every third node, but none of the current theories sheds any light on the change of phyllotaxy from alternate (2/5) to distichous with the attainment of the adult phase, or on the cyclical variations in internode length, diaphragm thickness or length of summer laterals which are found within the primary shoot. In the future, it is likely that greater emphasis will be given to experimental morphology, to the clarification of grapevine structure by scanning electron microscopy and to the role of phytohormones and growth regulators in determining phyllotaxy and the mode of construction of the stem (Srinivasan and Mullins, 1981a).

VASCULAR ANATOMY OF THE STEM

CONNECTION OF LEAVES TO THE PRIMARY SHOOT

Detailed studies of relationships between the vasculature of the leaf and stem of the eight youngest nodes of the elongating vine shoot have been made by Fournioux and Bessis (1973) using sections and cleared specimens. Procambial cells are initiated at the base of the leaf initials, the cells that give rise to the leaf primordia, and differentiation of procambium proceeds in the acropetal direction from the sub-apical zone into the leaves, where it gives rise to the petiolar vascular bundles. In the grapevine there are five traces per leaf, a large median trace and two pairs of lateral traces (Fig. 3.12). The connection of these traces to the vascular cylinder follows the distichous phyllotaxy of the primary shoot. Fournioux and Bessis (1973) have shown that the five traces that

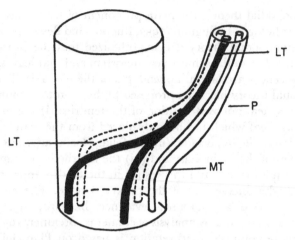

Fig. 3.12. The vascular connection between the stem and leaf consists of five traces (per leaf): a large median trace (MT) and two pairs
of lateral traces (LT). Within the node the lateral traces from each side
of the stem become joined but the median trace remains separate.
Three leaf traces enter the petiole (P). From Fournioux (1982). Reproduced with permission

serve each leaf are distinct for four internodes in the basipetal direction;
however, they join the vascular cylinder at that lower level. They pass
into the node holding the leaf at the upper level. Each trace enters the
node through a separate leaf gap. Within the node the two lateral traces
from each side of the stem become joined but the median trace remains
separate. Three leaf traces enter the petiole.

DEVELOPMENTAL ASPECTS OF VASCULAR ANATOMY

The shoot apical meristem has two functions: the production of new
organs and the production of new tissues. A rapidly growing shoot of the
grapevine increases in length by three to four centimeters per day and
produces a new leaf (or tendril) primordium every two to three days.
The histological changes which accompany extension growth and shoot
maturation in the grapevine have been summarized in diagrammatic
form (Fig. 3.13) by Fournioux (1982). In this diagram the anatomy of
the stem is shown at each of seven stages. Stage one represents the apical
meristem. In stage two the newly-produced cells undergo differentiation
to become either parenchymatous cells of differing types or procambial
cells which give rise to the primary vascular tissues. In stage three the
first derivative of the procambium, the protophloem, is produced cen-

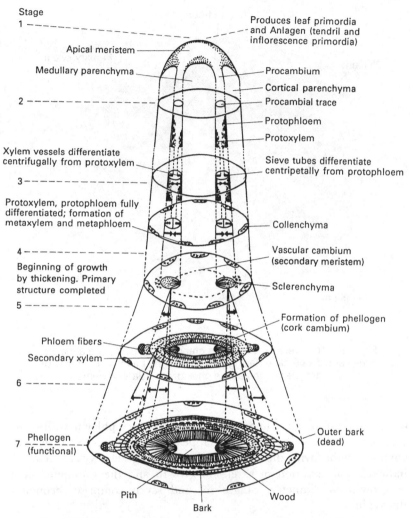

Fig. 3.13. Developmental anatomy of the grapevine cane. From an original drawing by Dr. J.C. Fournioux, Universite de Dijon. Reproduced with permission

tripetally. Protoxylem arises later and is formed centrifugally. The differentiation of protophloem and protoxylem is completed by stage four, but metaxylem and metaphloem are still largely undifferentiated; the metaxylem at this stage is composed of large unlignified parenchymatous cells. Collenchyma first appears at stage four. At the next stage (five) the differentiation of collenchyma is complete, as is the

(a)

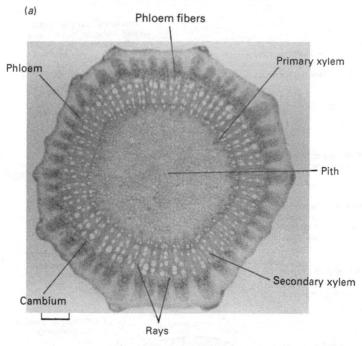

Phloem fibers

Phloem

Primary xylem

Pith

Secondary xylem

Cambium

Rays

Fig. 3.14. Transverse sections of a green stem (above) and a mature one-year-old vine cane (facing page). Note the prominent bundles of phloem fibers in the above photograph. (Bars = 1 mm.)

differentiation of the primary xylem and primary phloem within the vascular bundles (Fig. 3.13). The appearance of thick-walled scleren-chyma together with the presence of strongly lignified xylem elements indicates the cessation of extension growth and the commencement of growth in diameter. Stages six and seven comprise secondary thickening.

SECONDARY VASCULAR ANATOMY OF THE STEM

After differentiation of the primary vascular tissues, the cells towards the center of each vascular bundle enter into division to produce a cambial layer known as intrafascicular cambium. Secondary phloem is then pro-duced by this cambium in the direction of the epidermis, and secondary xylem is produced in the direction of the pith. Later, an interfascicular cambium is formed between the vascular bundles, and this too produces phloem to the outside and xylem to the inside. Together, the intra-

(b)

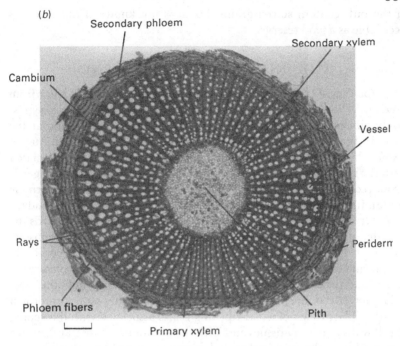

Secondary phloem

Secondary xylem

Cambium

Vessel

Rays

Periderm

Phloem fibers

Pith

Primary xylem

Fig. 3.14. (*cont.*)

fascicular cambium and the interfascicular cambium comprise the vascular cambium. The secondary phloem consists of two tissues: (i) soft phloem composed of sieve tubes, companion cells, parenchyma cells and phloem fibers; and (ii) hard phloem composed of thick-walled fibers. Hard phloem and soft phloem are formed in alternating bands, and this gives the bark a ringed appearance when viewed in transverse section (Fig. 3.14).

The secondary xylem is diffuse-porous (as distinct from ring-porous) and the vessels have scalariform or ladder-like thickening. These vessels are surrounded by living cells of xylem parenchyma. The predominant elements of the xylem are thick-walled septate fibers with bordered pits and living protoplasts. There are prominent rays. Medullary rays, those which separate the original primary vascular bundles, and rays of the second order, those which were initiated subsequently by the interfascicular cambium, traverse both the xylem and phloem and are widest towards the epidermis. Ray cells of both types contain chloroplasts, starch, crystals, tannin bodies and other inclusions. Xylem ray cells are lignified, but phloem ray cells have only the primary cell walls. Cells of the pith contain starch grains, but it is not known if this starch is accessible as a food reserve.

THE 'RIPENING' OF THE CANE

The ripening of the cane (aoûtement) refers to its change in color from green to yellow and thence to brown. This color change occurs progressively along the cane during the growing season, commencing at the base. In France the process is usually completed in the month of August (Août), hence the term 'aoûtement'. The anatomical changes associated with the ripening of canes are as follows. A cork cambium or phellogen is produced at the shoot base, 20 or more internodes basal to the internode containing the current transition to vascular cambium and secondary vascular tissue differentiation. Cork or phellem is produced towards the outside, and a phelloderm or 'cork-skin' is formed towards the inside. Together, phellogen, phellem and phelloderm are known as periderm. The formation of periderm results in the separation of the epidermis, cortex and primary phloem from the secondary phloem and other vascular elements of the main axis; these separated tissues soon turn brown and die. The process of wood-ripening is also associated with thickening of cell walls in the ray tissues and with the accumulation of starch grains (amyloplasts), the principal carbohydrate reserve of the grapevine, by all living cells of the wood and bark.

RADIAL GROWTH OF THE TRUNK AND ARMS

In cultivated grapes the trunk is trained to bear two or more main branches known as arms, which bear the fruiting canes. In practice the grapevine is pruned annually and the dormant one-year-old canes are cut back either to 'spurs' bearing two or three buds or to canes ('rods') bearing a dozen or more buds. Dormant canes are seldom left intact. Depending on the circumstances the arms are pruned to a given number of spurs or rods and provision is made for periodic replacement or renewal of arms. In each growing season there is thickening of the trunk and the arms and of the spurs or rods. This involves renewed activity of phloem, cambium and phellogen.

A characteristic of the phloem of the grapevine is that sieve tubes can remain functional for several seasons. At the end of the first growing season there is a heavy deposition of callose on the sieve plates. This callose is removed through the activity of glucan hydrolases at the beginning of the next season and the functioning of the sieve tubes is restored. Phloem in the trunk and arms can remain functional for three to

four years, and up to six alternating bands of hard and soft phloem can
be produced within a single growing season.

Renewed division of the cambium gives rise to new phloem to the
outside and new xylem to the inside. A single ring of xylem is produced
each year; the vessels of the early wood are larger than those of the late
wood. Vessels can remain functional for up to seven years, but most are
inactivated by tyloses when two to three years of age.

A new phellogen is formed each year from cells of the secondary
phloem. The phellogen arises in an inner layer of soft phloem of the
previous season. This secondary phloem is then isolated from the internal tissues of the stem through the production of phelloderm and
phellem, and it dies. This dead tissue is added to the accumulated
periderm of earlier years, which in due course sloughs off at the surface
as strips or flakes of dead bark. The production of periderm, together
with the effects of pruning, give the trunk and arms their characteristi-
cally rough and gnarled appearance. In long-lived grapevines the trunk
can achieve a substantial size; Galet (1976) refers to a vine in Portugal in
which the circumference of the trunk is two meters.

The root system

Much of the world's viticulture is found in Mediterranean or semi-arid
climates and is based on soils of inherently low fertility which, in addition, are often highly calcareous or salinized. The root system of the
grapevine must cope with water stress, waterlogging, ionic imbalance
and toxic ions. It is probable that the hardiness of grapevines resides to a
large extent in characteristics of the root system such as tolerance to
anoxia, capacity to penetrate the profile to a depth of three meters or
more (Champagnol, 1984), ability to regenerate new roots and to store
organic nutrients including amino acids (Nassar and Kliewer, 1966),
and presence of mycorrhizal associations (Possingham and Groot-
Obbinck, 1971).

The form of the root system is affected by the mode of propagation.
Plants grown from seed tend to have a tap-root system with a major axis,
which gives rise to secondary roots. Commercial viticulture is based on
clonal cultivars and plants are propagated by cuttings or by grafting
onto rootstocks. In this case the root system is more highly divided and
there are several classes of roots (Richards, 1983). The main framework
roots (6–100 mm diameter) are usually found at a depth of 30–35 cm
from the soil surface and their number is said to remain constant after
the third year from planting. Smaller, permanent roots (2–6 mm diameter) arise from this framework; they either grow horizontally, in which

case they are known as 'spreaders', or they may be 'sinkers' and grow downwards. These roots undergo repeated branching to produce the fibrous or absorbing roots, which are ephemeral and are continually being replaced by new lateral roots. Although sinker roots do penetrate to considerable depths, fibrous root, which accounts for the major portion of the root mass, is found primarily within the top 20–50 cm, depending on the cultivar, the soil type and the age of the plant (Champagnol, 1984).

ROOT MORPHOLOGY

The extremity of the young root bears a prominent root cap, which protects the root apex from damage as the elongating root pushes its way through the soil. The cap is continually being worn away at its tip and replaced by divisions of cells in the root tip. As in the roots of other species, the root cap of the grapevine is probably an organ of perception of gravity.

The root tip is two to four millimeters long and comprises the root apical meristem. Proximal to the tip for a distance of several millimeters is the zone of elongation, and this is followed by the zone of absorption of water and mineral ions (also known as the piliferous layer). This zone, which is whitish in color, is about 100 mm in length and is covered with root hairs, that is, epidermal cells that have elongated in a direction perpendicular to the root surface. Depending on the cultivar and the environment, root hairs achieve a density of 300–400 mm^{-2}; individual hairs are 12–15 μm in diameter and 140–365 μm in length. Root hairs are ephemeral and they are replaced by outgrowth of new hairs towards the root tip as they die towards the proximal end of the absorption zone. In this way a band of root hairs at a uniform distance from the tip is maintained continuously in the absorption zone of the elongating root. Proximal to the absorption zone the root has a brown color due to the presence of suberized phellem.

ROOT ANATOMY

The newly formed tissue close to the root tip consists of concentric layers of cortical parenchyma surrounding the procambial cylinder. The outer layer of this cortex constitutes the epidermis, the site of root hair production. Towards the interior many of the cortical cells contain tannin bodies or raphides of calcium oxalate. The innermost layer of the cortex becomes the endodermis. Vesicles and arbuscules of endomycorrhizal fungi are often found in the cortex.

At the exterior of the primary vascular cylinder is the pericycle, a tissue composed of three layers of cells. Within the pericycle there are the xylem bundles (two to eight) which alternate with phloem bundles. In transverse sections these bundles look like the spokes of a wheel. The primary phloem develops centrifugally and consists of sieve tubes and phloem parenchyma. Companion cells are absent. The primary xylem develops centripetally and contains a mixture of large and small vessels. Medullary or interfascicular rays are formed between the vascular bundles; lateral root primordia originate in the pericycle opposite the primary xylem.

In young white roots, suberization of the endodermis commences within 10 mm of the root tip. In this process, suberin lamellae are deposited between the cell wall and the plasmalemma of the endodermal cells surrounding the stele (Richards and Considine, 1981). Young roots of grapevines may be either white or brown. The brown color results from the oxidation of phenols released from vacuoles of dead or collapsed epidermal cells. When this occurs, the next layer of cells, the hypodermis, may develop suberin lamellae inside its cell walls. The factors that regulate this 'peripheral suberization' are not well understood. In grapevine roots, suberization is most extensive in mid-summer when the soil temperature is high and the soil water content is low. In dry soil some roots become suberized and brown to their tips. When soil conditions become more favorable the suberized roots may resume extension growth from the original root apical meristem or they may produce new lateral roots. It is probable that suberization of the hypodermis enables the primary roots to survive during periods of water stress or other unfavorable conditions.

Secondary thickening commences with the initiation of cambium within the primary phloem bundles; this cambial layer subsequently develops to encircle the primary xylem and soon the young secondary xylem. The vascular cambium so produced gives rise to secondary xylem towards the interior and secondary phloem towards the exterior. The vessels of the secondary xylem have scalariform thickening and are generally larger than those of the stem. The secondary phloem consists primarily of soft phloem. Phloem fibers are few in number. The rays are well developed and cell walls are thinner than in ray tissue of the stem. The rays, together with other living cells of the vascular system of the root, are sites of starch accumulation (Fig. 3.15).

A phellogen or cork cambium is formed from the innermost layer of the pericycle at the time of formation of the vascular cambium. As in the shoot, the phellogen produces cork towards the exterior and phelloderm towards the interior. The cortical tissues outside the corky layer turn

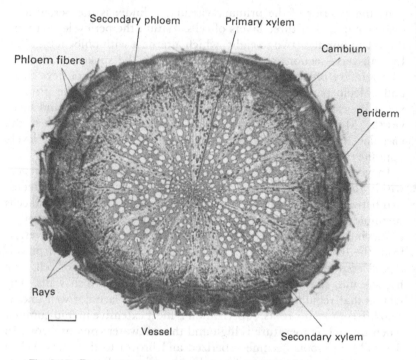

Fig. 3.15. Transverse section of a one-year-old grapevine root, showing extensive secondary thickening. (Bar = 1 mm.)

brown and die, and eventually they are sloughed off. In the following
seasons a new phellogen is formed within the secondary phloem of the
preceding season; the activity of this phellogen, like that of phellogen
formed in stems, gives rise to an annual shedding of bark.

The vascular cambium of the root is persistent and the age of grapevines can be determined by counting the annual rings of secondary
xylem. According to Galet (1976) the distinctness of the rings is enhanced by slight displacements from year to year in the orientation of
the rays and by differences in the appearance of early wood and late
wood within an annual ring.

Grapevine leaves

STRUCTURE

The newly formed leaf consists of the lamina, the petiole and a pair of
stipules. The latter structures lack leaf traces and are ephemeral. The

vascular supply of the lamina and the petiole is described in detail by Fournioux and Bessis (1977). Within the lamina the traces undergo repeated branching to give the characteristic palmate venation of the grapevine leaf. The leaf margins are toothed; each tooth ends in a hydathode, a water-excreting gland. The upper epidermis is almost devoid of stomata but bears a well-differentiated layer of epicuticular wax. This wax consists of platelets of so-called 'soft wax' (Radler and Horn, 1965) and differs in chemical composition from the epicuticular waxes of fruits, which contain both hard and soft waxes. The palisade tissue consists of a single layer of elongated cells in which the ratio of length to diameter is about 4 : 1. The palisade tissue accounts for up to 50% of the leaf thickness. The spongy mesophyll has five or six layers of polygonal or irregularly lobed cells and a very extensive system of air spaces. Mesophyll of both types is amply supplied with chloroplasts. The lower epidermis has many stomata and there are large substomatal cavities (see Fig. 4.1). The lower epidermis lacks a cuticle but it has hairs of various types (woolly, spiny, glandular) depending on the genotype. These hairs are usually found on the veins, and the hairiness of the underside of the leaf is a useful character for cultivar identification.

The leaf of the grapevine has two abscission zones, one at the junction of the lamina and the petiole and the other at the point of attachment of the petiole to the stem (Fig. 3.10). In the process of leaf fall in autumn, these abscission zones are activated by signals from the senescing lamina. In some cultivars the distal abscission zone is the first to be activated, and the laminae fall off in advance of the petioles.

AMPELOGRAPHY

Leaf morphology in the grapevine and its relatives is of special significance because differences among genotypes in leaf shape are the foundation of *ampelography*, the art and science of grapevine cultivar description and identification. In the genus *Vitis* the predominant leaf form is palmate, in which all of the main veins arise from a single point. There are five main veins, and these serve the five lobes of the leaf. Each of these main veins is numbered, and measurements of the angles between veins and the lengths of veins are encoded for use in keys for species and cultivar identification. Other components of the key are derived from measurements or observations on leaf outline (leaves are described as cordiform, cuneiform, truncate, orbicular or veniform), leaf size (product of length × width) and extent of lobes and sinuses (the spaces between the lobes are the sinuses); the shape of the petiolar sinus is of special importance. Records are also made on the extent and character of the leaf hair (cobwebby, downy, felty), leaf color, leaf contour (flat, folded, contorted, concave, convex) and on dentation (ratio of length to width of the teeth at the margin of the lamina).

In the ampelographic procedures developed by Galet (1971) descriptions of each genotype are produced in the form of a drawing of the leaf outline and a concise statement, which incorporates both quantitative and qualitative information. Other criteria of importance in ampelography are the shape and hairiness of the shoot tip, the characteristics of the herbaceous shoot, cane, and bunch, and the growth habit.

Anatomical aspects of grapevine propagation

ADVENTITIOUS ROOT FORMATION

Cultivars of *Vitis vinifera* are easily propagated by hardwood or softwood cuttings but several species of North American origin, notably *Vitis berlandieri*, are difficult to root by conventional procedures. Most grapevine cultivars, *vinifera* or hybrid, and most rootstocks are amenable to micropropagation and this procedure is of importance both for plant multiplication and for production of pathogen-free stock.

In both cuttings and explants there is production of callus at the base of the stem, but adventitious roots seldom arise from cells of the basal callus. Adventitious root primordia arise primarily from or near to the interfascicular cambium of both nodes and internodes. There are no pre-formed root initials in *Vitis vinifera*.

GRAFTING

In green grafting, the joining of herbaceous shoots of rootstock and scion, it is said that both the cambium and the pith participate in the healing process and in the formation of a functional union, but in mature one-year-old wood graft-healing is due to cambial activity alone. In both the stock and the scion, the surface cells of the bark and wood dry out and die. According to Pratt (1974) the vessels adjacent to the cut surfaces become sealed by tyloses and gum, and cambial activity commences close to the bud on the scion. Callus is produced, which differentiates to form periderm, cortex and vascular strands. There appears to be little or no information on the finer details of graft-healing in the grapevine, but it is unlikely that the process is different from graft-healing in other woody perennials of horticultural importance (Hartmann and Kester, 1976).

SEXUAL PROPAGATION

INFLORESCENCES AND FLOWERS

Flowering in the mature grapevine is normally a three-step process. The first step is the formation of *Anlagen* (singular *Anlage*) or uncommitted primordia by the apices of latent buds on shoots of the current season. Next, the Anlagen develop either as inflorescence primordia or as tendril primordia and shortly thereafter the latent buds enter into dormancy. In some circumstances Anlagen produce shoots instead of tendrils or inflorescences. Finally, the formation of flowers from the inflorescence primordia occurs at the time of bud burst in the next season.

Occasionally, grapevines growing in temperate environments produce inflorescences and flowers in the same season; this 'second crop' arises primarily from the summer lateral shoots. Some genotypes have a greater tendency than others to second cropping. When grapevines of temperate origin are grown in the humid tropics (South India, Thailand, Indonesia) they behave as evergreens, and inflorescences and flowers are formed on lateral shoots on each of the flushes of shoots which regrow after pruning. Such grapevines routinely produce two crops of fruit per year.

This chapter is concerned with temperate viticulture and with inflorescences borne on shoots that emerge in spring from dormant latent buds. The origin of inflorescence primordia in latent buds, and the subsequent development of the inflorescence and flowers, was studied by Srinivasan and Mullins (1976, 1981*b*) using scanning electron microscopy (Fig. 3.16). They proposed a developmental code in which the various stages (0–11) are related to changes in the shape of organs or to the addition of new structures. In this code the latent bud apex is considered to be vegetative until the appearance of the first Anlage, and buds containing only leaf primordia are classified as Stage 0.

FORMATION OF INFLORESCENCES AND FLOWERS

Depending on the cultivar, the latent bud apex produces three to eight leaf primordia (Stage 0) and then divides into two almost equal parts. The part opposite the youngest leaf primordium is the Anlage (Stage 1). The formation of Anlagen from the apex is the earliest indication of reproductive growth in the grapevine, and the formation of the Anlage can be regarded as the stage of initiation of the inflorescence axis. The first Anlagen are found about midsummer in latent buds at the base of the cane on the newly ripened wood. Thereafter, Anlagen appear

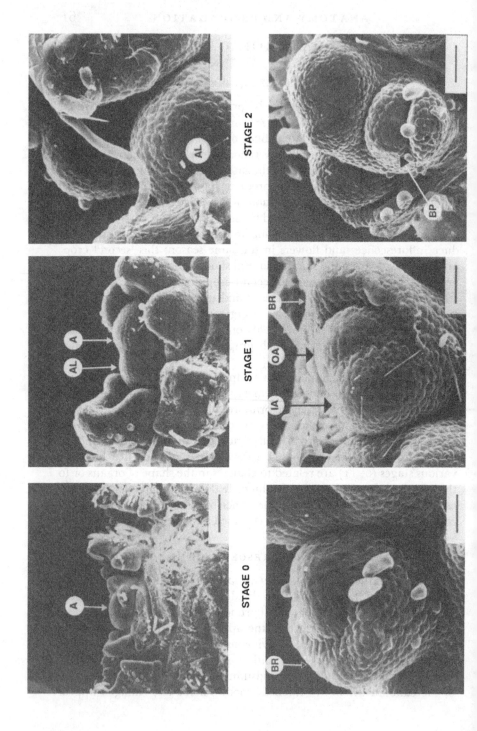

STAGE 0

STAGE 1

STAGE 2

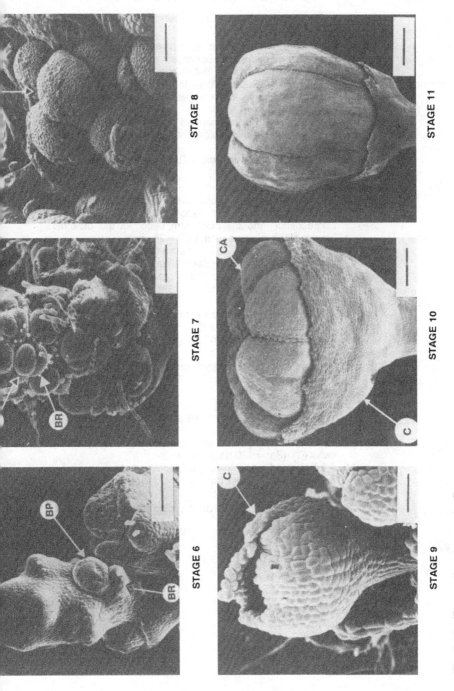

STAGE 8

STAGE 11

STAGE 7

STAGE 10

STAGE 6

STAGE 9

Fig. 3.16. (For caption see overleaf)

Fig. 3.16. Development stages in the flowering of *Vitis vinifera* L.

Formation of Anlagen

Stage 0. Depending on the cultivar, the apex (A) of a young latent bud forms a specific number of leaf primordia (5 in Shiraz) before the initiation of the first Anlage. (Bar = 85 µm.)

Stage 1. Bisection of the apex (A) to form the Anlage (AL). The Anlage is opposite the youngest leaf primordium. (Bar = 43 µm.)

Stage 2. The Anlage (AL) has separated from the apex and is developing into a blunt, broad, obovate structure. (Bar = 19 µm.)

Stage 3. Formation of bract primordium (BR) from the abaxial flank of the Anlage. (Bar = 19 µm.)

Formation of inflorescence primordia

Stage 4. Division of the Anlage to form an inner Arm (IA) and an outer arm (OA). The inner arm becomes the main axis of the inflorescence and the outer arm becomes the proximal branch of the inflorescence. (Bar = 26 µm.)

Stage 5. Growth of the main axis (inner arm) to give rise to the first branch primordium (BP). (Bar = 26 µm.)

Stage 6. Growth of the main axis of the inflorescence primordium to form several branch primordia (BP) and bract primordia (BR). (Bar = 43 µm.)

Stage 7. A fully developed inflorescence primordium in a mature dormant latent bud. Note the branch primordia (BP) and bracts (BR). (Bar = 88 µm.)

Formation of flowers

Stage 8. Differentiation of the branch primordia at bud burst and formation of the flower initials (FI). Note group are flower initials. (Bar = 43 µm.)

Stage 9. Development of calyx (C) in a flower initial (after bud burst). The calyx forms an incomplete cover over the developing flower. (Bar = 65 µm.)

Stage 10. The calyptra (CA) lobes are visible through the top of the calyx (C). (Bar = 130 µm.)

Stage 11. A fully formed grape flower just before anthesis. There is full development of flower parts: calyx, calyptra, stamens and pistils. (Bar = 250 µm.)

From Srinivasan and Mullins (1981*b*). Reproduced with permission progressively in latent buds towards the growing tip; their formation coincides with the change in color of the stem from green to brown (aoûtement).

Initially, the two parts of the divided apex (Anlage and latent bud apex) have a similar appearance (Stage 1) but they soon acquire different conformations. Anlagen develop as broad, blunt, obovate structures and they lack stipular scales (Stage 2). Leaf primordia are narrow pointed structures possessing stipular scales and they arise from the flank of the apex (Stages 0, 1).

The further development of the Anlagen starts with the formation of a bract. Bracts originate as depressions in the distal ends of the Anlagen. Later, these depressions appear to move to the periphery and to form a collar-like structure (Stage 3). The Anlage then divides into two unequal parts, called arms. The larger, adaxial part (nearer to the apex) is the inner arm, and the smaller, abaxial part (adjoining the bract) is the outer arm (Stage 4).

Inflorescence primordia are formed by extensive branching of the Anlage (Stage 5). The inner arm divides and produces several globular branch primordia, which give rise to the main body of the inflorescence. Branching of the outer arm is less extensive; it develops into the lowest branch of the inflorescence. The branch primordia of the inner and outer arms give rise to branch primordia of the second and third order, each of which is subtended by a bract (Stages 6 and 7). The degree of branching of the inner arm gradually decreases in an acropetal direction, giving the inflorescence primordium a conical shape (Stage 7). The appearance of a fully developed inflorescence primordium is rather like a bunch of grapes, in which each berry-like branch primordium is a protuberance of undifferentiated meristematic tissue (Stage 7). After the formation of one to three inflorescence primordia (depending on the cultivar), the latent bud enters dormancy. The differentiation of flowers from the inflorescence primordia begins after the dormant latent buds are activated in the spring.

Each branch primordium of the inflorescence primordium divides many times; ultimately, it produces the flower initials (Stage 8). There are differences among cultivars in regard to the differentiation of flowers. In Muscat of Alexandria and Thompson Seedless the flower primordia are formed in groups of three, but in Shiraz the flowers occur in groups of five.

Initiation and development of flower parts occurs simultaneously in all parts of the inflorescence primordium. Within each flower the sepals, petals (calyptra), stamens and pistils develop one after the other (Stages 9, 10, 11).

Fig. 3.17. The sequence of events from early bloom to maturation of the fruit. The photos depict approximately 20% (a) and 80% (b) bloom, dehiscence of calyptrae (c, d), pollination (e), fruit set (f, g) and mature bunch of grapes (h). See also Fig. 4.7. ((a, b, g, h), Bar = 20 mm; (c–f), bar = 2 mm.)

The appearance of the calyx as a continuous ring of tissue on the rim of the flower primordium marks the beginning of the formation of flower parts. The calyx ring does not coalesce at its tip but forms an incomplete cover over the petals (Stage 9). The petals and stamens arise as five papillae soon after the formation of sepals. The petals, which develop at the same time as the calyx, become lobed and then push their way through the calyx. Special cells are formed at the margins of petals, which interlock with similar cells on the margins of the adjacent petals to form a calyptra or cap (see Fig. 3.16). Anthesis or cap-fall in the grapevine usually occurs in the forenoon, and is triggered by changes in the turgor of the interlocking cells. As the temperature rises the petals become free at their bases; they then separate along the margins and

Fig. 3.17. (*cont.*)

curve upwards to release the stamens. The sequence of events from full bloom to dehiscence of calyptrae, fruit set and production of a ripe bunch of grapes is illustrated in Fig. 3.17.

FLOWER TYPE

Most species in the genus *Vitis* are dioecious (male and female flowers occur on different plants), but nearly all commercially important cultivars of *Vitis vinifera* are hermaphrodite: their flowers contain both functional stamens and functional pistils. According to Negi and Olmo (1971) the sexual type is determined by three alleles. The primitive hermaphrodite is $Su^+ Su^+$. A dominant mutation, Su^F, suppresses ovary development to produce maleness. In the presence of a recessive allele, Su^M, the filaments are reflexed, the pollen is sterile, and the flowers are functionally female. The dominance relations of these three alleles are

$Su^F > Su^+ > Su^M$. The hermaphrodite *vinifera* grapes are thus $Su^+ Su^+$ or $Su^+ Su^M$. A few female cultivars ($Su^M Su^M$) are of commercial importance, notably Ohanez and Katta Kourgan.

REPRODUCTIVE ANATOMY

GYNOECIUM

The ovary of the grapevine is superior; there are two locules, each containing two ovules with axile placentation. A detailed account of the development and histochemistry of the grape pistil is given by Considine and Knox (1979a). In brief, the style is short and the stigma is of the wet type. The tissues of style and stigma are rich in crystals of calcium oxalate. The stigmatic surface is covered with filamentous papillae; each papilla comprises about 20 cells. Mature receptive stigmas have a wet, glistening appearance on the day of anthesis; the stigmatic exudate is continuous with similar material in the transmitting tissue.

According to Considine and Knox (1979a) the mature ovary wall or pericarp consists of three tissues: (i) an outer epidermis; (ii) a mesocarp comprising inner and outer cortical parenchyma cells separated by a net of anastomosing vascular bundles (the main branches of these bundles are the ventral and dorsal carpellary traces) (Fig. 3.18); and (iii) an endocarp composed of crystal containing cells and inner epidermal cells.

At the base of the ovary is the disc, a whorl of nectaries which may or may not be functional. The nectaries are thought to produce the characteristic odor of the grapevine inflorescence but production of nectar is disputed on anatomical grounds.

OVULE AND EMBRYO SAC

The ovule of *Vitis* is anatropous, that is, inverted with the micropyle pointing towards the pedicel and with the funiculus (ovule stalk) joined to the outer integument forming a raphe (Fig. 3.19). There are two integuments, inner and outer, and a well-developed nucellus.

A single vascular strand traverses the funiculus and terminates at the chalaza. The outer integument is joined on one side to the raphe for most of its length but it is free at the micropylar end. The outer layer of the outer integument is rich in tannin bodies. The inner integument, which is composed of two or three layers, sometimes extends beyond the outer integument to form a collar-like structure or endostome around the micropyle.

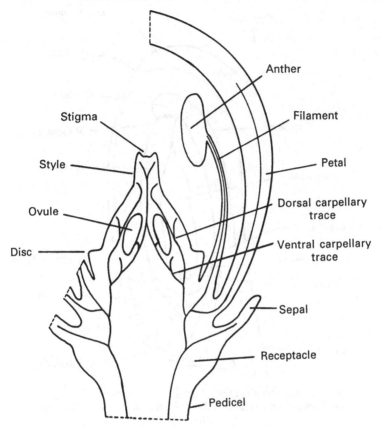

Fig. 3.18. Diagram of a longitudinal section of a mature flower. From Pratt (1971). Reproduced with permission

Nucellus, the tissue which surrounds the embryo sac, is extensive in the *Vitaceae*, but nucellar embryony *in vivo* has not been established. The nucellus tissue of *vinifera* grapes possesses a high degree of regenerative competence *in vitro*. Another characteristic of grapevine nucellus is the well-developed hypostase, a densely-staining, thick-walled tissue close to the chalaza.

The sequence of events leading to embryo sac formation is as follows. First, a subepidermal cell of the nucellus becomes an archesporial cell and it divides periclinally to produce an outer primary parietal cell and an inner primary sporogenous cell. The primary parietal cell gives rise to (i) the nucellar calotte, a tissue which lies between the embryo sac and the micropyle; and (ii) the nucellar cap, a 'plug' of cells between the two inner integuments close to the micropyle (Fig. 3.20*a*). The primary

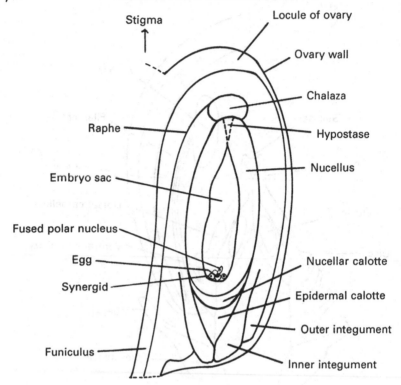

Fig. 3.19. Longitudinal section of an anatropous ovule of the grape cultivar Concord (at full bloom). From Pratt (1971). Reproduced with permission

sporogenous cell becomes the megaspore mother cell, which then undergoes meiosis to produce four haploid megaspores.

The megaspores are arranged initially in a linear tetrad. The megaspore towards the chalazal end of the ovule forms the embryo sac and the remaining three megaspores degenerate. There then follow three mitotic divisions to produce an eight-nucleate embryo sac of the normal or Polygonum type. In the grapevine the fusion of the polar nuclei to form the secondary nucleus occurs before fertilization and the antipodals are short-lived.

ANDROECIUM AND POLLEN FORMATION

The grapevine has five stamens, each consisting of a bilobed anther borne on a slender filament. Within each anther lobe are two pollen sacs

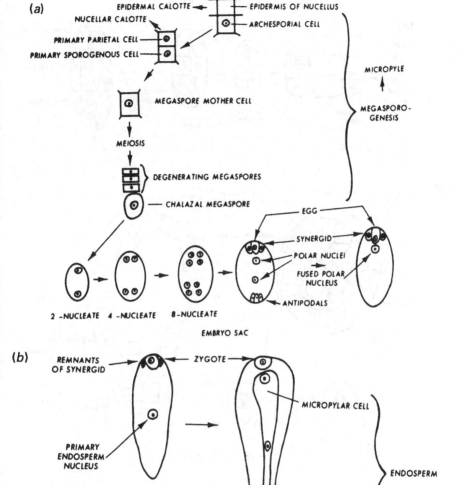

Fig. 3.20. (a) Embryo sac development. (b) Endosperm development (according to Helobial type). (c) Endosperm development (according to the Geum variation of the Asterad type). The terminal cells and its derivatives are stippled. The organs into which the embryonic regions will develop are indicated by arrows in the 18-celled stage. (d) Grape seeds. (A) Transverse section of a seed above the level of the embryo about 4 weeks after pollination. (B) Longitudinal section of a mature 'Concord' seed. (C, D) Surface view of 'Concord' seeds. From Pratt (1971). Reproduced with permission

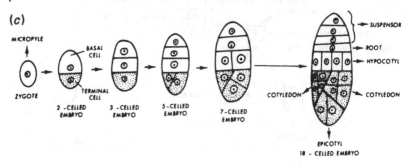

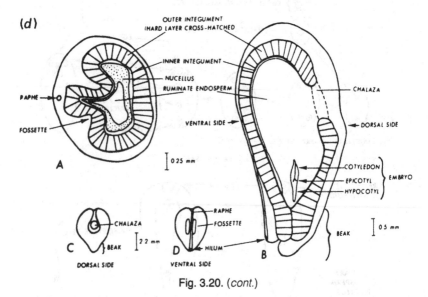

Fig. 3.20. (cont.)

or microsporangia. According to Pratt (1971) little is known of the early stages of microsporogenesis in *Vitis vinifera*, and the archesporium and its derivatives, the primary parietal layer and the primary sporogenous layer, are not well described (Fig. 3.21). The anther wall arises from the primary parietal layer and consists of three tissues, epidermis, endothecium (a tissue characterized by thick walls) and tapetum (a tissue characterized by varying numbers of nuclei per cell). The primary sporogenous layer produces the pollen mother cells (PMC); meiosis in each PMC gives rise to four haploid microspores. The tetrad stage and the newly released microspores are readily observed by light microscopy using standard staining procedures (e.g. Toluidine Blue following hydrolysis). Thereafter, observation of grapevine pollen by light micros-

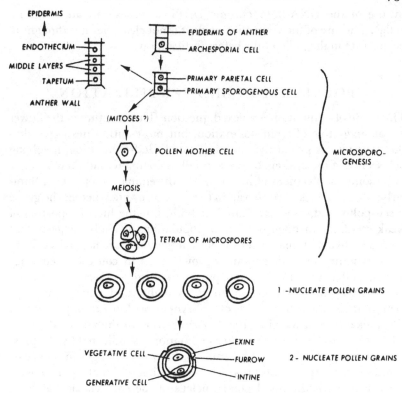

Fig. 3.21. Pollen development in the grapevine. Redrawn from Pratt
(1971)

copy is made difficult by the formation of a thick exine. The microspores
of *vinifera* cultivars are released into the locule in a uninucleate condition.
At this stage the locule contains anther sap (tapetal periplasmodium),
a nutrient fluid formed by degeneration of the tapetum and which
is present up to the time of anthesis. The first pollen grain mitosis
occurs synchronously in microspores shortly after their release into the
locule. The generative nucleus, the vegetative nucleus, and the wall that
separates them are often difficult to resolve owing to the thick exine.
Mature pollen grains often contain starch, as evidenced by I–KI staining. Grapevine pollen is tricolpate – it has three furrows each with a
germ pore – and the pattern of dehiscence of anthers is intorse, that is,
they open towards the centre of the flower by means of a longitudinal
split.

Microgametogenesis occurs by mitotic division of the generative
nucleus within the pollen tube; the two sperms are readily observed

by use of the DNA-fluorochrome DAPI (Rajasekaran and Mullins, 1983). The fate of the vegetative nucleus is not clear, but it is thought to degenerate in the pollen tube before fertilization.

POLLINATION AND FERTILIZATION

The mode of pollination is a vexed question. The structure of the flower is not suggestive of wind-pollination, but most authorities agree that the grapevine is primarily wind pollinated. Bud-pollination, involving dehiscence of the anthers before cap-fall, is common, and there is likely to be some involvement of insects in the dissemination of pollen. Similarly, there is a lack of consensus in the literature as to whether the grape is self-pollinated, cross-pollinated or both, but the high proportion of weak seedlings in open-pollinated populations strongly suggests that *vinifera* cultivars, unlike their dioecious progenitors, are normally selfed.

In experiments with container-grown vines in controlled environments, pollen germination and pollen tube growth were favored by high temperatures (27 °C day to 22 °C night) and pollen tubes appeared at the micropylar ends of embryosacs within 12h of pollination (Rajasekaran and Mullins, 1985). Fertilization in the vineyard generally occurs two or three days after pollination, as indicated by the presence of the pollen tube at the embryo sac or by changes in ovules or ovaries. Pratt (1974) notes that 'the association of a male gamete with the egg or with the fused polar nucleus of the embryo sac has been observed rarely or not at all.'

EMBRYO, ENDOSPERM AND SEED

The embryology of the grapevine has not been much studied, and information on embryo, endosperm and seed formation is fragmentary.

The pattern of embryo formation in the grapevine is classified as the Geum variation of the Asterad type (Maheshwari, 1950) (Fig. 3.20c). The zygote undergoes an unequal division to produce a small terminal cell towards the chalaza and a large basal cell towards the micropyle. Derivatives of the terminal cell give rise to the apical meristem (gemmule) and cotyledons and derivatives of the basal cell give rise to the hypocotyl, the primary root and the suspensor, a tissue which attaches the embryo to the wall of the embryo sac. The pattern of endosperm formation in the grapevine is classified as helobial (Maheshwari, 1950). The primary endosperm nucleus divides and a transverse wall is formed across the embryo sac, forming a small chalazal cell and a large micropylar cell (Fig. 3.20b). Within the micropylar cell the nucleus enters into free divisions; up to six divisions occur in the absence of wall formation. Within the chalazal cell every division is accompanied by wall formation. In the mature seed the endosperm is irregular in shape and is composed mainly of small thick-walled cells containing oil, aleurone grains or crystals.

After fertilization there is a period of rapid cell division in the funiculus, raphe, chalaza and integuments. At the micropyle the outer integument thickens and elongates to form the beak (Fig. 3.20d). Towards the chalaza the outer integument becomes folded and a ridge is produced at the raphe on either side of which are two depressions or grooves known as fossettes. The pincer-like growth of the outer integument squeezes the nucellus and endosperm into a W-shape (Fig. 3.20d). The endosperm continues to grow within the rigid case formed by the outer integument, and by 35 days after anthesis it has crushed and replaced the nucellus.

The hardening of the grape seed is due to lignification of the inner layers of the outer integument. This hard layer is thick at the beak and thin at the fossettes. The inner integument remains thin and adheres to the endosperm. Grape seeds are extremely resistant to decay and are of considerable archeological interest. The morphology of the mature seed provides useful criteria for identification of cultivars and species and the characteristics of fossilized or carbonized seeds, particularly the degree of development of the beak, have been much studied in relation to the paleobotany of *Vitis* and the history of viticulture.

SEEDLESSNESS

Several well-known cultivars of grapevines produce seedless fruit. There are two main types of seedlessness: (i) parthenocarpy, in which the berries develop without fertilization and are entirely seedless; and (ii) stenospermocarpy, in which the berries each contain one or more aborted seeds (Stout, 1936). Parthenocarpy is exemplified by the cultivar Zante Currant, in which there is degeneration of the embryo sac. Pollination alone is the trigger for fruit development, but fruit set is enhanced by girdling or by growth regulator treatment. The cultivar Sultanina (syn. Thompson Seedless) is stenospermocarpic. Pollination and fertilization are triggers for fruit development but there is a failure of endosperm and embryo development. The integuments contain little or no sclerenchyma and the aborted ovules appear as small soft whitish seeds in the mature fruit. Ovule development in stenospermocarpic genotypes may be normal or near normal; seedlings of seedless grapes can be produced *in vitro* by culture of fertilized ovules containing rudimentary zygotic embryos (Cain *et al.*, 1983). The pollen of stenospermic cultivars is usually functional, and seedlessness is heritable.

ANATOMY OF THE BERRY

There is much variation in the literature in the naming of the component tissues of the grape berry. According to Pratt (1971) the fruit wall,
from its outer surface to its inner surface adjacent to the seed, is the
pericarp and has five components: epidermis, hypodermis, outer wall,
inner wall and inner epidermis. Viala and Péchoutre (1910) refer to
the epidermis as epicarp, the middle of the wall as mesocarp and the
inner epidermis as endocarp. Fournioux (1982) also refers to a pericarp
with three main components: (i) epicarp or pellicule comprising cuticle,
epidermis and hypodermis; (ii) the mesocarp or pulp; and (iii) the
endocarp or inner wall of the pulp.

Considine and Knox (1979b, 1981) describe the skin of the grape
berry as the dermal system comprising the outer epidermis of the
pericarp, its covering of wax and cuticle and the collenchymatous
hypodermis. They stress the importance of the dermal system because it
supports and protects the fruit, and because its constituent tissues contain a high proportion of the substances responsible for pigmentation,
flavor and aroma. Considine (1981a,b) has also investigated the fine
structure of the dermal system in relation to fruit splitting, a problem
associated with rainfall during fruit maturation.

The characteristic bloom of the grape berry is composed of overlapping platelets of cuticular or epicuticular wax. These hydrophobic
platelets help to prevent the loss of water from the fruit (Possingham et
al., 1967). Chemically, epicuticular wax consists of two thirds hard wax,
mainly oleanolic acid, and one third soft wax. Soft wax is a complex
mixture of organic compounds, but fatty alcohols are the major components (Radler and Horn, 1965). The cells of the mesocarp, pulp, inner
and outer walls of the pericarp (according to which nomenclature is
preferred) constitute the flesh of the berry. The cells towards the skin
and the inner epidermis are rounded in outline and tend to be smaller
than the elongated cells at the center of the pericarp. The septum
grows to fill any locule where seeds have aborted. The growth of the
grape berry and changes in pericarp cells will be discussed further in
Chapter 5.

VASCULAR ANATOMY OF THE BERRY

The pedicel of the developing flower has five or six bundles, which separate in the receptacle to give branches that serve the flower parts and
branches that serve the ovary. In the developing fruit the remnants of
the flower-bundles can be seen in the 'bourrelet' (the remains of the
receptacle). The ovary-bundles give rise to a complex network of vascu-

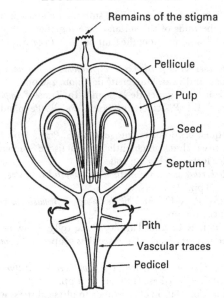

Fig. 3.22. Vascular anatomy of the fruit. Diagram of a longitudinal section of a young berry showing the course of vascular bundles. Modified from Bouard (1971)

lar traces within the berry. This network has three components, two of which, the bundles serving the seeds and the bundles serving the placenta, have a common origin in the septum. This vascular tissue and its associated parenchyma is known colloquially as 'the brush' in English and 'le pinceau' in French; it is the material that remains attached to the end of the pedicel when a ripe grape is plucked from the bunch. The third component is the highly ramified peripheral or superficial bundles, which are located at the junction of epicarp and mesocarp, i.e. at the junction of the outer wall and inner wall of the pericarp. The peripheral bunches are joined to the central bundles (Fig. 3.22).

Literature cited

Alleweldt, G. 1963. Einfluss von Klimafactoren auf die Zahl der Infloreszenzen bei Reben. *Wein Wissensch.* **18**: 61–70.
Alleweldt, G. and Balkema, G.H. 1965. Uber die Anlage von Infloreszenz- und Blutenprimordium in den Winterknospen der Rebe. *Z. Acker- u. Pflbau* **123**: 59–74.
Antcliff, A.J., Webster, W.J. and May, P. 1957. Studies on the Sultana vine: V. Further studies on the course of bud burst with reference to time of pruning. *Aust. J. Agric. Res.* **8**: 15–23.

Barnard, C. 1932. Fruit bud studies. I. The sultana: an analysis of the distribution and behaviour of the buds of the sultana vine together with an account of the differentiation and development of the fruit buds. *J. Coun. Sci. Industr. Res. Aust.* **5**: 47–52.

Bessis, R. 1965. *Recherches sur la fertilité et les correlations de croissance entre bourgeons chez la vigne.* Thèse, Docteur d'Etat, Université de Dijon. 236 pp.

Bouard, J. 1971. Tissus et organes de la vigne. In *Traité d'Ampélologie: Sciences et techniques de la vigne.* (Ed. P. Ribéreau-Gayon and R. Peynaud), vol. 1, p. 121. Dunod, Paris.

Branas, J. 1957. Sur quelques données ontogénétiques. *Progr. Agric. Vitic.* **148**: 48–67.

Breidner, H. 1953. Entwicklungsgeschichtlich – genetische Studien über somatische Mutationén bei der Rebe. *Züchter* **23**: 208–22.

Bugnon, F. 1953. *Recherches sur la ramification des Ampélidacées.* Presses Universitaires de France, Paris. 158 pp.

Bugnon, F. and Bessis, R. 1968. *Biologie de la vigne: Acquisitions récentes et problèmes actuels.* Masson. Paris. 160 pp.

Cain, D.W., Emershad, R.L. and Tarailo, R.E. 1983. In-ovulo embryo culture and seedling development of seeded and seedless grapes (*Vitis vinifera* L.). *Vitis* **22**: 9–14.

Champagnol, F. 1984. *Eléments de physiologie de la vigne et de viticulture générale.* Francois Champagnol, Saint-Gely-du-Fesc (France). 351 pp.

Considine, J.A. 1981a. Correlation of resistance to physical stress with fine structure in the grape, *Vitis vinifera* L. *Aust. J. Bot.* **29**: 475–82.

Considine, J.A. 1981b. Stereological analysis of the dermal system of fruit of the grape *Vitis vinifera* L. *Aust. J. Bot.* **29**: 436–74.

Considine, J.A. and Knox, R.B. 1979a. Development and histochemistry of the pistil of the grape, *Vitis vinifera. Ann. Bot.* **43**: 11–22.

Considine, J.A. and Knox, R.B. 1979b. Development and histochemistry of the cells, cell walls and cuticle of the dermal system of the fruit of the grape *Vitis vinifera* L. *Protoplasma* **99**: 347–65.

Considine, J.A. and Knox, R.B. 1981. Tissue origins, cell lineages and patterns of cell division in the developing dermal system of the fruit of *Vitis vinifera* L. *Planta* **151**: 403–12.

Eichler, A.W. 1878. *Blütendiagramme*, part 2. W. Engelmann, Leipzig. (Cited by May, P. 1964.)

Fournioux, J.-C. 1982. *Sciences de la vigne: Cours de travaux pratiques.* Université de Dijon.

Fournioux, J.-C. and Bessis, R. 1973. Etude du parcours caulinaire des faisceaux conducteurs foliaires permettant la mise en évidence d'une rythmicité anatomique chez la vigne (*Vitis vinifera* L.). *Rev. Gén. Bot.* **80**: 177–85.

Fournioux, J.-C. and Bessis, R. 1977. Principales etapes de l'histogenèse vasculaire dans les nervures de la feuille de vigne (*Vitis vinifera* L.). *Rev. Gén. Bot.* **84**: 377–95.

Galet, P. 1971. *Précis d'ampélographie pratique.* Third edition. P. Galet, Montpellier. pp. 266. (English translation and adaptation by L.T. Morton, 1979. Cornell University Press, Ithaca and London. 248 pp.)

Galet, P. 1976. *Précis de viticulture.* Third Edition. P. Galet, Montpellier. 586 pp.

Hartmann, H.T. and Kester, D.E. 1976. *Plant propagation. Principles and practices.* Third Edition. Prentice-Hall, Englewood Cliffs, New Jersey. 662 pp.

Kasimatis, A.N. 1982. Annual growth cycle of a grapevine. In: Grape pest management (ed. D.L. Flaherty, F.L. Jensen, A.N. Kasimatis, H. Kido and W.J. Moller),

Cooperative Extension, University of California, Division of Agriculture and Natural Resources, Publication 4105, pp. 12–29.

Maheshwari, P. 1950. *An introduction to the embryology of angiosperms*. McGraw-Hill Book Company, New York, 453 pp.

Mullins, M.G., Nair, Y. and Sampet, P. 1979. Rejuvenation *in vitro*: Induction of juvenile characters in an adult clone of *Vitis vinifera*. *Ann. Bot.* **44**: 623–8.

Nassar, A.R. and Kliewer, H.M. 1966. Free amino acids in various parts of *Vitis vinifera* at different stages of development. *Proc. Amer. Soc. Hort. Sci.* **89**: 281–94.

Negi, S.S. and Olmo, H.P. 1971. Conversion and determination of sex in *Vitis vinifera* L. *Vitis* **9**: 265–79.

Pearson, R.C. and Goheen, A.C. 1988. *Compendium of grape diseases*. APS Press, St. Paul, Minnesota. 93 pp.

Plantefol, L. 1949. A new theory of phyllotaxis. *Nature* **163**: 331–2.

Possingham, J.V., Chambers, T.C., Radler, F. and Grncarevic, M. 1967. Cuticular transpiration and wax structure and composition of leaves and fruit of *Vitis vinifera*. *Aust. J. Biol. Sci.* **20**: 1149–53.

Possingham, J.V. and Groot-Obbink, J. 1971. Endotrophic mycorrhiza and the nutrition of grapevines. *Vitis* **10**: 120–30.

Pratt, C. 1971. Reproductive anatomy in cultivated grapes. A review. *Am. J. Enol. Vitic.* **22**: 92–109.

Pratt, C. 1974. Vegetative anatomy of cultivated grapes. A review. *Am. J. Enol. Vitic.* **25**: 131–50.

Radler, F. and Horn, D.H.S. 1965. The composition of grape cuticle wax. *Aust. J. Chem.* **18**: 1059–69.

Rajasekaran, K. and Mullins, M.G. 1983. The origin of embryos and plantlets from cultured anthers of hybrid grapevines. *Am. J. Enol. Vitic.* **34**: 108–13.

Rajasekaran, K. and Mullins, M.G. 1985. Somatic embryo formation by cultured ovules of Cabernet Sauvignon grape: Effects of fertilization and of the male gameticide toluidine blue. *Vitis* **24**:151–7.

Richards, D. 1983. The grape root system. *Hort. Reviews* **5**: 127–68.

Richards, D. and Considine, J.A. 1981. Suberization and the browning of grapevine roots. In *Structure and function of plant roots*. (ed. R. Brouwer, O. Gasprikova, J. Kolek and B.C. Loughman, pp. 111–15. Martinus Nijhoff/Dr. W. Junk, The Hague.

Snyder, J.C. 1933. Flower bud formation in the Concord grape. *Bot. Gaz.* **94**: 771–9.

Srinivasan, C. and Mullins, M.G. 1976. Reproductive anatomy of the grapevine (*Vitis vinifera* L.): Origin and development of the Anlage and its derivatives. *Ann. Bot.* **38**: 1079–84.

Srinivasan, C. and Mullins, M.G. 1981a. Modification of leaf formation by cytokinin and chlormequat (CCC) in *Vitis*. *Ann. Bot.* **48**: 529–34.

Srinivasan, C. and Mullins, M.G. 1981b. Physiology of flowering in the grapevine. A review. *Am. J. Enol. Vitic.* **32**: 47–63.

Stout, A.B. 1936. Seedlessness in grapes. *N.Y. Agr. Exp. Stn. Tech. Bull.* **238**, 68 pp.

Tucker, S.C. and Hoefert, L.L. 1968. Ontogeny of the tendril in *Vitis vinifera*. *Am. J. Bot.* **55**: 1110–19.

Viala, P. and Péchoutre, F. 1910. Morphologie du genre *Vitis*. In: Viala, P. and Verrnorel, V. *Traité général de Viticulture*. (*Ampélographie* **1**: 113–90.) Masson, Paris.

4

Developmental physiology: the vegetative grapevine

Carbohydrate production by the vine

PHOTOSYNTHESIS

The woody skeleton of a deciduous perennial such as the grapevine is a storehouse for reserves, and it provides both carbohydrates and mineral nutrients for growth during the early part of the season following bud burst. Once an appreciable leaf surface has been established by the emerging shoots, photosynthesis is the primary source of carbon for growth and for replenishment of reserves.

Plants are unique in that they are able to utilize energy from solar radiation and convert it into chemical energy in order to reduce CO_2 from the atmosphere to produce carbohydrates. The organelle responsible for this conversion is the chloroplast. It is composed of a double outer membrane, which envelops the stroma and the internal lamellar membranes. The stroma contains the soluble enzymes of metabolism, starch storage and protein synthesis. The enzyme present in highest concentration in the stroma is ribulose1,5-bisphosphate carboxylase–oxygenase (RuBPC/O), and it may constitute up to 50% of the total protein within the chloroplast. This enzyme catalyzes the reaction by which CO_2 is incorporated into the photosynthetic carbon reduction pathway. The lamellar membranes, which individually are called thylakoids, contain chlorophyll and are involved in the biophysical reactions of solar radiation capture and its conversion into usable forms of energy.

The processes involved in photosynthesis can be separated into the 'light' and 'dark' reactions. The light reactions are concerned with the conversion of light energy into chemical energy in the forms of adenosine triphosphate (ATP) and nicotinamide adenine dinucleotide phosphate (NADPH). The chlorophyll molecules within the chloroplast capture light energy and transfer it to photochemical reaction centers known as photosystems. The carotenoid pigments (carotenes and xanthophylls) also assist in capturing light energy in regions of the spectra not absorbed by chlorophyll. It is the flow of electrons through the two photosystems (termed photosystems I and II) that result in the formation of ATP and NADPH needed for the assimilation of CO_2. These electrons come from the splitting of water molecules in photosystem II, and molecular oxygen (O_2) is a byproduct of this reaction.

The dark reactions of photosynthesis, known as the photosynthetic carbon reduction cycle (PCR cycle or Calvin cycle), utilize the energy captured by the light reactions for the reduction of CO_2 to carbohydrates. The 'dark' reactions can occur only while there is a supply of ATP and NADPH. However, the supply of ATP and NADPH generated through the light reactions in a leaf is sufficient to maintain the PCR cycle for only a very short time (less than a few seconds) after a plant is put into darkness. Therefore, the PCR cycle operates in very close conjunction with the light reactions.

The first step in the PCR cycle involves the incorporation of CO_2 by the carboxylation of ribulose phosphate by RuBPC/O. The first stable products of this reaction are two molecules of glycerate 3-phosphate (glycerate 3-P). Plants in which the initial carboxylation reaction result in the formation of three carbon acids are termed C_3 plants. All *Vitis* species are C_3 plants. The cycle continues when glycerate 3-P is phosphorylated (when ATP is present) to form glycerate 1,3-P. This molecule is then reduced (when NADPH is present) to form glyceraldehyde 3-P. This triose phosphate, together with dihydroxyacetone phosphate (DHAP), can be combined to form fructose 1,6-bisphosphate (fructose-P_2). Fructose 6-P is formed through the action of fructose bisphosphatase (FBPase). This enzyme regulates the flow of carbon within the PCR cycle and in several other pathways inside the chloroplast. Fructose 6-phosphate can be used either to form starch through several intermediate steps or for the continuation of the PCR cycle. The PCR cycle is completed when ribulose 5-P and ATP, in the presence of ribulose 5-P kinase, form ribulose-P_2. In C_3 plants, 2 NADPH and 3 ATP are needed for the conversion of CO_2 into carbohydrates.

There are several factors that control the flow of carbon within the PCR cycle, most of which involve the regulation of the enzymes of the cycle. Several of the enzymes within the cycle are regulated by light. This is mediated by changing conditions, within the chloroplast, which alter the enzymes. In addition, changes in the stromal pH and magnesium concentration, due to changes in going from darkness to light, affect the activities of both RuBPC/O and FBPase.

As mentioned earlier, RuBPC/O also functions as an oxygenase. Both O_2 and CO_2 compete for RUBP at the catalytic site on the enzyme. Therefore, net photosynthesis of the leaf is directly inhibited by O_2, owing to competitive inhibition of RuBPC/O by O_2 with respect to CO_2. The oxygenase reaction also results in the formation of phosphoglycolate, which is hydrolyzed in the chloroplast to glycolate. P-glycolate is the starting point for the photosynthetic oxidative carbon cycle (C_2-cycle), which is the metabolic pathway for 'photorespiration' (Husic *et al.*, 1987). Photorespiration is the term used to denote the uptake of O_2 and release of CO_2 that occurs in the light in photosynthesizing organisms. The release of CO_2 from this pathway has been estimated to be equivalent to 15–50% of the rate of net CO_2 assimilation (Sharkey, 1988). The release of CO_2 in photorespiration has been viewed as a wasteful process, but the O_2 cycle may provide (i) a means to convert undesirable products of photosynthesis, P-glycolate and glycolate, to carbohydrate as efficiently as possible and (ii) a mechanism to protect reaction centers of photosynthesis against photooxidation during periods of water stress and high irradiance (Osmond, 1981; Powles, 1984).

STOMATA

The physical movement of CO_2 into and out of the grape leaf is illustrated in Fig. 4.1. Carbon dioxide diffuses down a concentration gradi-

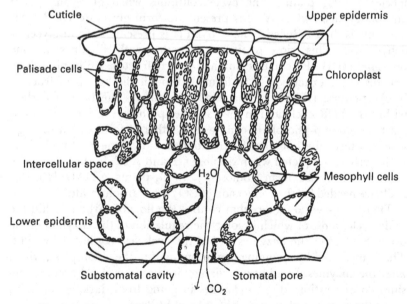

Fig. 4.1. Diagrammatic representation of a cross-section of a grape leaf showing the pathway for diffusion of water vapor out of the leaf and of CO_2 into the leaf

ent from the ambient air, through the stomata, into the leaf and then to the chloroplasts. Grape leaves are hypostomatous, that is, they have stomata only on the lower surface. The frequency and distribution of stomata varies with environmental conditions, stage of growth and genotype, but the average frequency is approximately 170 stomata per square millimeter. Since the flux of CO_2 through the leaf cuticle is minimal, the diffusion of CO_2 through the stomata and the control of stomatal functioning directly affect the rate of leaf photosynthesis. Light causes stomata to open and darkness causes them to close. The stomata of grape leaves are fully open at one-tenth full sunlight (about 200 μmol $m^{-2} s^{-1}$). Other environmental factors that affect the functioning of stomata include CO_2 concentration, temperature, leaf-to-air vapor pressure difference, and pollutants. Water deficits also play a major role in controlling stomatal aperture and may override stomatal responses to other environmental factors. The water relations of the vine will be discussed later in this chapter.

ENVIRONMENTAL REGULATION OF PHOTOSYNTHESIS

Studies on photosynthesis in fruit plants have used two main approaches: (i) measurements of CO_2 uptake by individual leaves; and (ii) measurements of CO_2 uptake by whole plants. In discussing factors affecting photosynthesis, one should realize that photosynthesis of a single leaf may not always reflect the behavior of all the leaves within the grapevine's canopy. This is because leaves of different ages and positions within the canopy do not behave in the same way as mature leaves on the outside of the canopy, which are fully exposed to solar radiation.

The most important environmental factor controlling the rate of photosynthesis by single leaves in grapevines growing under optimal conditions is solar radiation. The light response curve for grapevine leaf photosynthesis is best described as a rectangular hyperbole. Light saturation for photosynthesis in C_3 plants is generally assumed to occur at one-third full sunlight (a photon flux density of 600–700 μmol $m^{-2} s^{-1}$). Any increase in solar radiation intensity above the saturation level will not result in an increase in leaf photosynthesis. However, the conditions under which vines are grown may cause some change in the value for light saturation (Kriedemann, 1968). For example, leaves on Perlette grapevines grown in the Coachella Valley of California, a desert climate, did not reach their maximum rate of photosynthesis until solar radiation reached a value of approximately 1500 μmol $m^{-2} s^{-1}$ (Fig. 4.2). This value is similar to that reported for field-grown Riesling vines

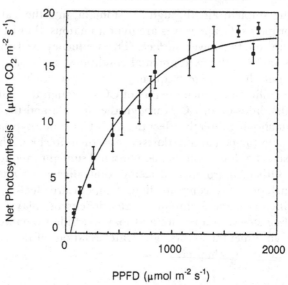

Fig. 4.2. The light response curve of net photosynthesis for leaves of field-grown Perlette grapevines in the Coachella Valley of California. Each value is the mean of three replicates ± one standard error (PPFD, photosynthetic photon flux density)

(Downton *et al*, 1987). The light compensation point for photosynthesis refers to the light level at which there is no net uptake of CO_2. At the compensation point, about 50 µmol m^{-2} s^{-1}, CO_2 that is assimilated by photosynthesis is equal to the loss of CO_2 from respiration. Approximately 90% of the solar radiation striking an individual leaf is absorbed by that leaf. Therefore, under conditions in which there are multiple layers of leaves in the canopy, leaves in the interior of the vine are shaded and are not photosynthesizing at maximum capacity. Such leaves may be at or below their light compensation point. Trellis type, vine density and row orientation play important roles in determining the efficiencies of solar radiation interception by a vineyard; these factors will be discussed in more detail in Chapter 6.

Temperature is another important environmental factor that determines the rate of photosynthesis in grape leaves. The optimum leaf temperature for photosynthesis of field-grown vines is quite broad; it is generally between 25 and 35 °C (Fig. 4.3) (Kriedemann, 1968). The rate of photosynthesis decreases to almost zero at 10 °C, and it also decreases at temperatures above the optimum. It should be noted that, even at leaf temperatures above 45 °C, rates of photosynthesis of field-grown vines

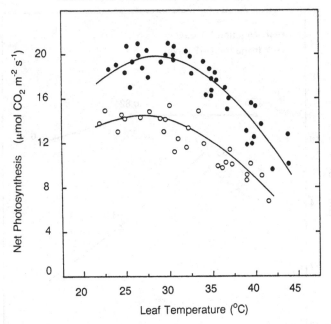

Fig. 4.3. The temperature response curve of net photosynthesis for leaves of field-grown Thompson Seedless grapevines in the San Joaquin Valley of California. The solid circles represent mature leaves with a nitrogen concentration of approximately 32.0 mg N g^{-1} dry mass. The open circles represent older leaves with a nitrogen concentration of approximately 24.0 mg N g^{-1} dry mass

can still be 50% of the rates measured at the optimum temperature (Fig. 4.3).

Mineral nutrients also can affect the capacity of a leaf to photosynthesize. The leaf's concentration (or content) of N is linearly related to its ability to fix CO_2 (Williams and Smith, 1985). This is expected because of the high concentration of RuBPC/O in the leaf, and because of the importance of this enzyme in regulating the PCR cycle. Very young grape leaves have high concentrations of N in their laminae (greater than 5% on a dry mass basis), but they do not have high rates of photosynthesis. This is due to the fact that the photosynthetic apparatus of young leaves is not fully developed. Therefore, the positive correlation between N and photosynthesis is evident only after the leaves are mature. The photosynthetic rate during leaf senescence will decrease owing to a decrease in N in the blade during senescence (Williams and Smith, 1985) (Fig. 4.3). Other mineral nutrients such as phosphorus (Skinner and Matthews, 1990) also affect the capacity of a leaf to fix CO_2.

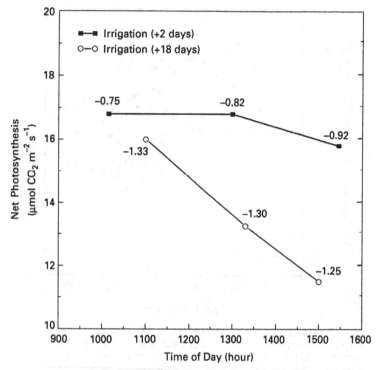

Fig. 4.4. Leaf photosynthesis of Colombard grapevines in response to irrigation frequency. Vines were irrigated either 2 (+2 days) or 18 (+18 days) before measurements. Values above each data point represent leaf water potentials (MPa) at the time of measurement. Redrawn from Downton *et al.* (1987)

EFFECTS OF CULTURAL PRACTICES ON PHOTOSYNTHESIS

Irrigation management is an important aspect in the culture of grape-vines where rainfall is inadequate during the seasonal growth of the vine. Irrigation affects the water relations of the vine, which in turn affect photosynthesis. Vines that have recently been irrigated will have rates of photosynthesis that can be maintained throughout the day (Fig. 4.4). This is in contrast to a longer interval between irrigations which will result in a decrease in photosynthesis from shortly before midday until sunset. The decrease in photosynthesis under these conditions is not due to a decrease in leaf water potential during the experimental portion of the day. Diurnal changes in ABA production (Loveys and During, 1984) may account for the observed differences in diurnal photosynthesis rates of well-watered and deficit irrigated vines. The work of Downton et al. (1987) and others indicated that deficit irrigation or infrequent irrigation can lead to considerable decreases in leaf photosynthesis which may be translated into lower yields.

The object of many cultural practices in viticulture is to increase fruit size or fruit quality. Size of berries is a prime consideration in table grape production because there is strong demand for large fruit, especially of seedless cultivars. Berry size in seedless grapes is increased by the use of gibberellic acid (GA_3) sprays at anthesis and berry set (approximately 10 days after anthesis), and by trunk girdling, removal of a strip of bark from around the trunk at the time of berry set.

Trunk girdling reduces the rate of leaf net CO_2 assimilation by about 25% during the four weeks after bark removal, but vines that have been girdled and sprayed with GA_3 at fruit set have photosynthetic rates which are between those of girdled vines and intact vines (Roper and Williams, 1989) (Fig. 4.5). The concentration of the phytohormone

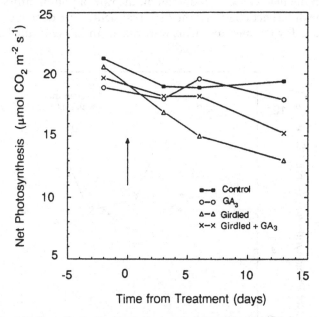

Fig. 4.5. The effect of trunk girdling and gibberellic acid (GA_3) sprays on the rate of net photosynthesis of Thompson Seedless grapevines. The arrow denotes day when these cultural practices were applied to the vines. There were significant differences among treatments 6 and 13 days after the treatments were imposed. Adapted from Harrell and Williams (1987)

abscisic acid (ABA) increases in leaves of girdled grapevines (During, 1978). ABA has long been implicated in controlling stomatal function-ing of plants (Raschke, 1979), and it appears that the reduction of photosynthesis in girdled vines is due to the effect of girdling on redu-cing stomatal aperture. Gibberellic acid sprays may overcome the effect of ABA on stomatal functioning and allow for stomata to remain open even when the vine is girdled.

During vineyard establishment entire shoots may be removed in order to form the vine for subsequent training. Once the vines are ma-ture, both leaves and shoots may be removed by summer pruning to improve fruit quality, to allow for ease of harvesting or to assist in obtaining better coverage of applied chemicals. There is some evidence that the photosynthetic rate of the remaining leaves after pruning may increase, owing to the change in the ratio of the carbon source (leaves) and the remaining sinks (importing organs). This phenomenon has been observed in vines with a single stem growing in a pot (Kriedemann *et al.*, 1975; Hofacker, 1978). Removing the fruit (sink) from Riesling vines grown in the field caused a reduction in the rate of photosynthesis (Fig. 4.6). The differences between the two treatments were dependent upon the time of day the measurements were taken. In this case, the lack of a

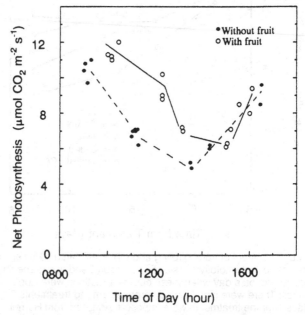

Fig. 4.6. The diurnal time course of net photosynthesis of Riesling vines grown in the field. Redrawn from Downton *et al.* (1987)

large sink resulted in a decrease in the rate of photosynthesis and not an increase in the rate of the vines with the fruit.

Carbohydrate utilization

RESPIRATION

The equation commonly used to summarize photosynthesis in green plants is:

$$n\,CO_2 + 2n\,H_2O \xrightarrow[\text{chlorophyll}]{\text{light+}} (CH_2O)_n + n\,O_2 + n\,H_2O.$$

The equation used to summarize the utilization of carbon substrates for the production of energy by living organisms is:

$$C_6H_{12}O_2 + 6\,O_2 \rightarrow 6\,CO_2 + 6\,H_2O + \text{energy}.$$

Respiration involves the catabolism of sugar or other carbon substrates with release of CO_2 and the consumption of O_2. Respiration occurs in the cell's cytoplasm and mitochondria; in biochemical terms, it includes glycolysis, the oxidative pentose phosphate pathway, the tricarboxylic acid cycle and the mitochondrial electron transport chain. During the process of respiration, low-energy bonds in the carbon substrates are converted to high-energy bonds in reduced nucleotides (NADH, NADPH, and $FADH_2$) and ATP. This energy conversion allows more thermodynamically difficult work to be accomplished. The oxidation of one glucose molecule has the potential to yield 36 ATP molecules. Another important aspect of the catabolism of these substrates is the production of carbon skeletons for use during the anabolic phase of metabolism within the plant.

The utilization of various substrates during respiratory processes influences the ratio of the moles of CO_2 produced per mole of O_2 consumed. This ratio is called the respiratory quotient (RQ). The RQ for the oxidation of glucose via respiration is unity. The RQ of lipids and proteins is less than one; the oxidation of organic acids is greater than one. Thus, it is possible to determine the nature of the substrate utilized during a particular stage of the vine's vegetative or reproductive growth by measuring the RQ of the tissue concerned. It should be noted, however, that partial oxidation of substrates tends to give anomalous results and RQ measurements require careful interpretation.

The major environmental factor regulating the rate of respiration is temperature. Increasing temperatures cause a progressive increase in respiration rate until the point at which tissue damage occurs. The respiration rate (CO_2 evolution) of a mature grape leaf is close to zero at a temperature of 10 °C. A typical rate of leaf respiration at a temperature of 25 °C is 0.5–1.0 μmol CO_2 m^{-2} s^{-1}. This is only a fraction of the maximum rate of net CO_2 assimilation by the leaf during photosynthesis. For many plants and their constituent organs, an increase in temperature of 10 °C will cause the rate of respiration to double, i.e. the Q_{10} of respiration is two. The Q_{10} of a particular organ may change throughout the growing season or it may change owing to differences in growth conditions.

Another important factor that controls the rate of respiration is the availability of substrates for oxidation. The respiration rate of an entire plant is not just a function of its size but is coupled to the daily assimilation of carbohydrates supplied by photosynthesis. This may not be true for woody perennial crops such as grapevines because the permanent structures of the plant contain a large supply of carbohydrate reserves, which may be utilized in the absence of a supply of photosynthate. The respiration rates of different organs of the plant will differ one from the other, as will the rate of an individual organ at different growth stages. The respiration rate of a meristematic region, such as the shoot and root apex or cambium, is much higher than that of fully differentiated tissue.

Respiration has been divided into two components: respiration required for growth, and respiration needed for organ maintenance. Growth respiration is associated with the biosynthesis of new phytomass, and its cost (carbohydrate requirement) is dependent upon the composition of the structure. Maintenance respiration in mature organs has an average cost of from 0.015 to 0.6 kg CO_2 kg^{-1} dry mass d^{-1}. Energy derived from respiration in mature tissues is used to meet the demands of many physiological processes, including carbohydrate translocation, protein turnover, nitrogen assimilation and ion uptake in the roots. Growth respiration on a whole-plant basis is greatest during periods of active growth, but the rate of maintenance respiration is dependent upon plant size. Although there have been no measurements in grapevines of the percentage of photosynthate utilized in respiration, estimates in other woody perennial species range from 38 to 65% (Kramer and Kozlowski, 1979). It is probable that respiration by the entire vine commands a large proportion of the daily photosynthate.

CARBOHYDRATE PARTITIONING WITHIN THE LEAF

Photosynthetic intermediates produced in the PCR cycle are used for starch synthesis in the chloroplast and for sucrose synthesis in the cell's cytoplasm. Regulation of starch and sucrose synthesis involves changes in the concentrations of various metabolites and inorganic phosphate, and in the pH optima of the various enzymes involved. Recently, it has been shown that fructose 2,6-bisphosphate (F2,6P) plays a major role in the regulation of carbohydrate metabolism in plants (Huber, 1986). To date, seven enzymes have been identified that are regulated by F2,6P, including those that direct photosynthate either to sucrose synthesis or to starch synthesis.

When sucrose accumulates, that is, when its rate of formation exceeds its rate of removal by the transport system, photosynthate is diverted into starch synthesis. The regulation of sucrose and starch formation is affected by environmental conditions and by changes in source–sink relations within the vine. Leaf temperature, for example, affects the relative concentrations of sucrose and starch within the leaves of grapevines (Buttrose and Hale, 1971). There are major diurnal changes in the starch content of grapevine leaves. Those that are fully exposed to solar radiation throughout the day have 1.5–2.5 times more starch just before sunset than at daybreak (Roper and Williams, 1989).

The rate of starch accumulation in leaves is a function of the length of the daily photosynthetic period and is not affected by changes in photoperiod (Chatterton and Silvius, 1979). There is little degradation of starch during periods of active photosynthesis; it is during the evening that the breakdown of starch occurs within the leaf. The initiation of starch degradation may be due to endohydrolytic action by hydrolases (Beck and Ziegler, 1989), but subsequent reactions which break the glycosidic bonds between the glucose units of starch involve phosphorolysis and are mediated by the enzyme, phosphorylase.

CARBOHYDRATE TRANSLOCATION

Sucrose is the main carbohydrate translocated throughout the grapevine. Generally, non-reducing sugars (of the raffinose series) represent the major if not sole carbohydrates that are transported in all higher plants. However, the mechanism by which carbohydrates move is still not fully understood. The theory of translocation proposed by Münch (1930) involved both diffusion and simple mass flow. Today, the driving force for movement of solutes is thought to involve a facilitated transfer mechanism (Giaquinta, 1983). Sucrose moves from the source of its synthesis through the mesophyll cells, and then enters the apoplast (free space) where it is loaded against a concentration gradient into a sieve element – companion cell complex (SE–CC) by co-transport with protons. Although the data favor apoplastic loading, recent attention has been given to an entirely symplastic route to the SE–CC within leaves (Turgeon, 1989). Accordingly, the unloading of solutes from the phloem may involve either the apoplastic or symplastic systems. The phloem forms a continuum in which solutes are transported by bulk flow at high rates. The active loading of sucrose and the resultant water flux across the phloem membranes appears to be sufficient to account for the observed rates of transport. The approximate range of rates observed in numerous studies is between 1 and 10 pmol cm^{-2} s^{-1} (Giaquinta, 1983). Regardless of the route, the loading of sucrose into and out of the phloem requires a source of energy. It is estimated that about 1.4% of the amount of sucrose that is earmarked for translocation would be needed to provide the ATP requirement for phloem loading (Giaquinta, 1983).

Translocation during the transition from a developing leaf to one that only exports its photosynthate is not fully understood. It appears that import by a leaf is terminated by some restriction of the phloem unloading pathway (Turgeon, 1989). The autoradiographic studies by Hale and Weaver (1963), in which different portions of the shoot were allowed to assimilate $^{14}CO_2$, indicate that there is a distinct pattern of photosynthate translocation within the shoot and that this pattern is dependent upon stage of shoot development. Photosynthate (^{14}C labeled) from mature leaves can be translocated either acropetally (to shoot tips and young leaves) or basipetally (to clusters or the permanent structures of the vines) depending upon location of the leaf on the shoot and its relation to the predominant sink at the time.

Vine growth

PHENOLOGY

Phenology is the study of events or growth stages of plants or animals that recur seasonally, and their relations with various climatic factors including temperature, solar radiation and day length. The aim of phenological studies is to describe or correlate the timing of specific growth stages with climatic factors or with other phenotypic events. Typically, the environmental factor may be in the form of a summation or some other transformation in which two or more variables are used. Degree days, growing degree days, day degrees and heat summations are variables that have been used to correlate ambient temperature with the plant's growth stage. Generally, the daily summations of temperature above a base temperature (10 °C) are calculated between the occurrence of one event and the next. Early use of degree days in California involved recording the mean of the daily maximum and minimum temperature and subtracting the base temperature. A major fault with this method of calculating degree days is that it does not take into account periods of fog, cloud cover or wind, factors that change ambient temperature during the period concerned, but which may have little effect on the daily maximum and minimum temperatures. With the advent of data loggers, it is now possible to calculate 'degree minutes,' and this overcomes the limitations of using daily minimum and maximum temperatures to calculate degree days (Williams, 1987c).

In grape production, phenological considerations are very important for selecting cultivars that will mature their fruit within a certain timeframe in the environment of the proposed vineyard location. Amerine and Winkler (1944) used accumulated degree-days above 50 °F (10 °C) to formulate recommendations for the growing of wine grape cultivars in California, where environments differ from cool coastal valleys and slopes to hot inland deserts. The five regions of Amerine and Winkler (1944), I (cool) to V (hot) have achieved wide acceptance, but more recent studies suggest that degree days may not be the most accurate basis for viticultural recommendations (McIntyre et al., 1982, 1987).

Knowledge of the various phenological stages of vine growth and their identification are important when performing various cultural practices or using chemical means to control either insects or pathogens. There have been several studies that have delineated, to varying degrees, the various stages of shoot and fruit growth (Baggiolini, 1952; Eichhorn and Lorenz, 1977; Pratt, 1971). The phenological events that are of importance in grape culture and pest control include: dormancy, budbreak, inflorescences clearly visible, anthesis or bloom (50–80% of the calyptras fallen), fruit set, veraison, fruit maturation, and leaf fall. An illustration of these phenological stages, according to the scheme of Eichhorn and Lorenz (1977), is shown in Fig. 4.7.

BUD DORMANCY AND BUDBREAK

The study of bud dormancy in deciduous woody perennials is a fertile field for semantics, but there is general agreement that latent buds of the grapevine exhibit three phases of dormancy (Huglin, 1958; Antcliff and May, 1961). The first phase is known as *conditional dormancy* and refers to the behavior of latent buds on green, herbaceous shoots early in the season. At this stage, buds do not normally grow out on the vine, but they burst readily if the shoot tip is removed, and even more readily if the leaves and lateral shoots are also removed. In herbaceous shoots of the grapevine the dormancy of latent buds is a manifestation of correlative inhibition.

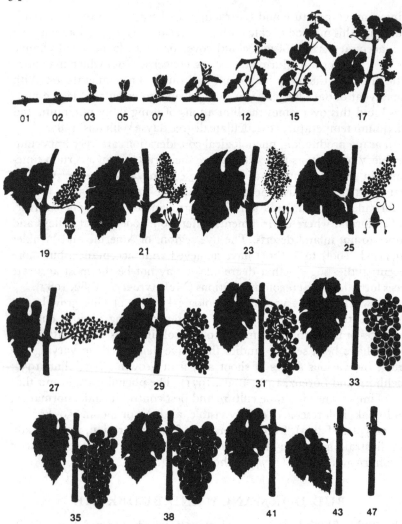

Fig. 4.7. Phenological stages in the growth and fruiting of grapevines. The more important phenological stages represented here are: bud swell (03), green shoot visible (05), clusters visible (12), bloom (19 to 25), set (27) and fruit maturity (38). From Eichhorn and Lorenz (1977). Reproduced with permission

The phases of dormancy in the grapevine are not distinctly separate but merge into one another and are separated by arbitrary divisions. For example, the second phase, *organic* or *deep dormancy*, develops progressively in the latent buds of ripened canes in late summer and early autumn. When cuttings are taken at regular intervals during this period and are forced under favorable temperatures, there is a progressive increase in the time required for bud burst. Organic dormancy commences when bud burst in cuttings takes more than 20 days, and it is at a maximum when buds do not burst in 70 to 75 days. Organic dormancy declines with the onset of winter. The final phase of dormancy is known as *enforced dormancy* or *rest* and is the state of buds towards the end of winter. At this stage, the only factor that prevents the outgrowth of buds is the low temperature in the vineyard. Cuttings that are brought into warm growing conditions burst their buds in only a few days.

There have been few studies on the endogenous control of bud dormancy in grapevines, but the involvement of abscisic acid has been proposed (Emmerson and Powell, 1978). Treatments that break organic dormancy and promote bud burst in cuttings include anaerobiosis, total immersion of cuttings in water at 30 °C for up to 72 h, and exposure to the vapor of ethylene chlorhydrin (Pouget and Rives, 1958). The dormancy of vine buds is prolonged by treatment with gibberellin (Weaver, 1959).

Many perennial tree crops require a certain number of 'chilling units' (hours below a certain temperature) before their buds will break. After chilling, buds will resume growth once environmental conditions become favorable. It is thought that grapevines require a certain minimum temperature for a given length of time in order to promote the breaking of organic dormancy in latent buds, but there is little quantitative information on chilling requirements.

In the humid tropics, the grapevine behaves as an evergreen, and shoot growth is continuous if vines are repeatedly pruned. In many temperate or subtropical regions with mild winters, budbreak is erratic and the percentage of buds that grow out is low compared with that of similar cultivars grown where winter temperatures are lower. Various chemicals have been used to promote bud burst, including hydrogen cyanamide (H_2CN_2). Applications shortly after pruning are very effective in producing an earlier and more uniform time of budbreak (Shulman et al., 1983). The use of cyanamide is particularly important in the production of early-market table grapes in subtropical environments. Another means of improving budbreak of grapes in areas where winters are characterized by sunny, clear days is through the use of evaporative cooling (Nir et al., 1988). The use of overhead sprinklers during winter decreases the temperature of buds exposed to direct sunlight, and this results in an earlier and more uniform budbreak. In temperate climates, budbreak occurs when the daily mean maximum temperature exceeds 10 °C, and models to predict this event have used temperature summations during the period following organic dormancy (Baldwin, 1966; Williams et al., 1985b). Recently, the base temperatures required for budbreak in dormant cuttings has been shown to vary among cultivars (Moncur et al., 1989). Lastly, some cultural practices from the preceding growing season seem to affect the date of budbreak, notably the use of deficit irrigation management in the vineyard, where water stress is applied during the season (Williams et al., 1991).

SHOOT AND LEAF GROWTH

Growth is defined as an irreversible increase in the size of the plant. It has two components: (i) an increase in the size of cells already present and (ii) an increase in the number of cells by divisions within meristems. Cell enlargement is due to water uptake that results from differences in the water potential between the cell and its surrounding components. The mechanical properties of the tissue's cells also play a role in the enlargement process. A loosening of the structural components of the cell wall must take place, and this must be localized so that cells in an apical meristem are able to elongate. Reactions include the breakage of covalent bonds within the cell wall or other changes to loosen the wall and apposition of new wall polymers. Water uptake, changes in mechanical properties of cells, and changes in biochemical processes all take place simultaneously during growth. Regulation of growth is through hormonal action, which, in turn, affects the constituent process of cell division and cell enlargement.

Under non-limiting conditions the growth of plants is described as exponential, at least during part of the growing season. With complex organisms such as grapevines, this type of growth in individual organs may not persist, owing to competition for carbohydrates from other organs, which results in a cessation of cell division and enlargement. In grapevines, the growth of the vegetative structures derived from compound buds is close to exponential early in the growing season (Fig. 4.8). Subsequent to anthesis, vegetative growth rates decrease and the growth curve for the shoots (leaves and main axis of the shoot or stem) of the vine under field conditions becomes sigmoidal (Fig. 4.8) (Alexander, 1958; Coombe, 1960; De La Harpe and Vissor, 1985; Williams, 1987a). This type of growth curve occurs whether the time variable is calendar days (De La Harpe and Vissor, 1985) or degree days greater than 10 °C (Gutierrez et al., 1985; Williams et al., 1985a; Williams, 1987a). The increase in dry mass of shoots is almost linear until fruit-set, when the majority of the dry mass increment of the vine is partitioned to the developing bunches.

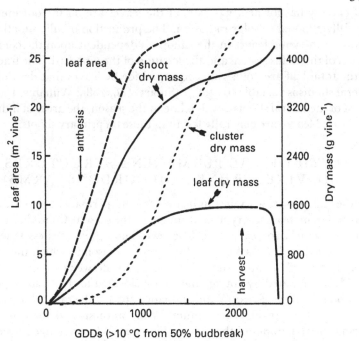

Fig. 4.8. The increase in dry mass of the leaves, shoots and clusters and leaf area of Thompson Seedless grapevines as a function of growing degree days (GDDs) greater than 10 °C. Data averaged over a three year period. Adapted from Williams and Matthews (1990)

Although most of the photosynthate is partitioned to the fruit after anthesis, the specific mass of the leaves (mass per unit area) and the stems (mass per unit length) still continues to increase (Williams, 1987a). The increase in mass per unit stem length is probably associated with the formation of periderm and with the accumulation of carbohydrates in the shoot. The increase in mass per unit leaf area is associated with an increase in structural components, probably of cell walls. The extent of this increase is positively correlated with the temperature of the leaf or canopy. The mass per unit leaf area has also been shown to increase under deficit irrigation (Williams and Grimes, 1987). Again, this may be due to temperature effects because the canopy temperature of water-stressed vines is greater than that of non-stressed vines.

The pattern of development of the vine's leaf canopy is similar to that of its shoots. The increase in leaf area is best described by a sigmoidal function (Williams, 1987a). Continued development of the canopy results in extensive mutual shading. It has been estimated that, once a full canopy has formed, 33–85% of the leaves are on the outside and are fully exposed to solar radiation. The proportion of fully functional, photosynthesizing leaves in the canopy is dependent upon the configuration of the trellis system and the location of the vineyard. The fraction of the total leaf area of a vine accounted for by leaves that develop on lateral shoots is variable (6–40%) (Smart *et al.*, 1985; Williams, 1987*a*). Most of the lateral shoots develop late in the season; the areas of individual lateral leaves are generally less than those on primary shoots.

GROWTH OF THE PERMANENT STRUCTURES OF THE VINE: TRUNK AND CORDONS (ARMS)

Studies on pot-grown vines indicate that the permanent structures of grapevines increase in dry mass throughout the season (Conradie, 1980). The trunks of two-year-old, field-grown Thompson Seedless vines almost tripled in dry mass during the growing season in which the trunk was established (Araujo and Williams, 1988). Between 10 and 30% of the ^{14}C assimilated by young vines is translocated to the trunk depending upon the time of year (Yang and Hori, 1979). The rate of increase in diameter of the trunk is maximum during anthesis and then declines during the remainder of the season (Fig. 4.9). When averaged over the life of a grapevine, 240 g dry mass vine^{-1} year^{-1} was partitioned to the

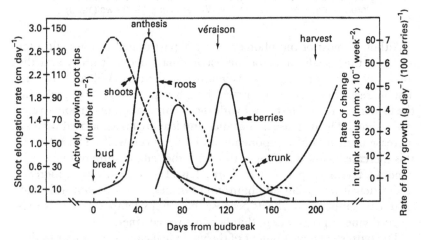

Fig. 4.9. The growth rate of shoots and trunk and actively growing roots in relation to the growth rate of the fruit of Colombard grapevines grown in South Africa. From Williams and Matthews (1990). Reproduced with permission

trunks of 18-year-old Cabernet Sauvignon grapevines grafted onto 5 C rootstock (Williams and Biscay, 1991). This is in contrast to an average of approximately 300 g dry mass vine^{-1} year^{-1} for trunk and cordons, respectively, of 12-year-old Chenin blanc vines grafted onto 101-14 Mgt rootstock (Saayman and van Huyssteen, 1980). These results indicate that the amount of carbon fixed by the vine and partitioned to the trunk and cordons varies throughout the growing season, with the age of vines and stage of vineyard establishment, and with the genotype.

ROOT GROWTH

There have been few quantitative studies of root growth in field-grown grapevines. Much labor is involved in extracting root systems from the soil, and measurements of root biomass are often inaccurate because of the loss of fine roots. Most studies of grapevine root systems have made use of underground root observation chambers to determine the periodicity of new root initiation and turnover (Hiroyasu, 1961; Freeman and Smart, 1976; McKenry, 1984; van Zyl, 1984). These studies have shown that a flush of root growth occurs shortly after shoot growth commences in spring and that it peaks at anthesis (Fig. 4.9). A second, major, flush of root growth begins after the fruit has been harvested. These two flushes of new root production arise from the permanent roots of the vine. New roots are white in color, but they eventually become suberized and turn brown. From these observations it appears that there is alternation of growth by aerial organs and root growth, and that root growth occurs only when an excess of photosynthate is available. However, when root biomass is used as a measure of growth, root dry mass increases throughout the entire period from shortly after budbreak until leaf fall for both young (Araujo and Williams, 1988) and mature vines (Williams and Biscay, 1991) (Table 4.1). The importance of flushes of root growth in relation to uptake of water and mineral nutrients is unknown.

DISTRIBUTION OF ROOTS IN THE SOIL PROFILE

The distribution of roots in the soil profile is influenced by edaphic characteristics and by cultural practices. The distribution of roots is affected by soil strength, compaction (natural and man-made) and by the presence of impervious layers. Cultural practices that affect root distribution include the type of irrigation system (drip or flood), vine density, and the rooting characteristics of the individual rootstock–scion combination. The majority of roots are found in the top one meter

Table 4.1. *The distribution of dry mass (grams per vine) among the organs of Chenin blanc grapevines grown in the San Joaquin Valley of California*[a]

| Organ | Date | | | |
	Mar. 19 (budbreak)	May 27 (1 week after anthesis)	July 25 (2 weeks after véraison)	Sept. 10 (fruit maturity)
Roots[b]	2139	2535	3020	2984
Trunk	2535	2870	2961	3015
Cordons	3319	3646	3284	3421
Stems	—	1143	1198	2539
Leaves[c]	—	1185	1469	1732
Clusters	—	519	2901	5199
Aerial:Root[d] ratio	2.74	2.57 (3.70)	2.07 (3.91)	2.16 (5.33)

[a] The vines were grown at the University of California Kearney Agricultural Center, near Fresno. The vines were planted in 1976 and entire vines harvested during the 1986 growing season. Vine and row spacings were both 2.4 m. The vines were trained to bilateral cordons at a height of 1.6 m. There were no foliage wires above the cordons.

[b] The roots were removed with a backhoe and were separated from the soil by hand. Roots were dug to a depth of 2.0 to 2.5 m. At this depth there was a hardpan that the roots were unable to penetrate.

[c] Total vine leaf areas for the vine harvests, beginning with May 27, were 23.2, 29.6 and 18.9 m^2 vine^{-1}, respectively.

[d] The aerial portion includes the trunk + cordons for the first row, and all aerial organs for the row in parenthesis.

of soil, but they can be found at depths of 6 m (Seguin, 1972). Under drip irrigation, approximately 70% of the total roots are in the top meter of soil, whereas under furrow irrigation there is a more even distribution of roots throughout the profile (Araujo, 1988). With drip irrigation, soil moisture may become depleted between the rows of vines as the season progresses, and the greatest volume of roots is found between vines, close to the points of watering.

The distribution of roots is also influenced by the cultivar and rootstock. The spatial distribution of roots tends to be dictated by soil type, but the density of roots within the profile is a function of rootstock (Southey and Archer, 1988). Williams and Smith (1991) found that the distribution of roots on deep, non-irrigated soil planted with Cabernet Sauvignon grafted onto Rupestris St George, 5 C or AxR # 1, was similar, but that the density of roots differed among the three combinations, the main factor being the rootstock. Similarly, Swanepoel and Southey (1989) found that, under intensive irrigation, rooting distribution and density of Chenin blanc was greatly affected by the rootstock. They also found that the above-ground growth (vegetative growth and yield) was greater as the root density and distribution increased.

FUNCTIONS OF THE ROOT SYSTEM

In addition to anchorage of the vine, a major function of the root system is the uptake of water. The transport of water to the root occurs readily if there is an abundance of soil moisture (soil water potential greater than −0.1 MPa), but it becomes progressively more difficult as the water content of the profile becomes depleted. This is exacerbated by the small proportion of the root system's surface that may be in contact with water within the soil (Atkinson, 1980). The greatest loss of water from a plant is via transpiration through the stomata. The loss of water from the leaf will lower its water potential. There are large diurnal fluctuations in vine leaf water potential under field conditions (Fig. 4.10). Well-watered vines have leaf water potentials close to zero just prior to dawn; the water potential declines during the day as the evaporative demand increases. Leaf water potential will recover subsequent to the midday minimum and approach predawn values late in the afternoon. Midday leaf water potential will generally decrease throughout the growing season; however, minimum values of vines irrigated with the amount of water they actually use should not be less than − 1.0 MPa even in semi-arid regions (Grimes and Williams, 1990). Vines growing in soils with less available water may undergo even greater diurnal variations in leaf water potential. The leaf water potential of vines grown under drought conditions will be more negative than those given adequate water on both a diurnal (Figure 4.10) and a seasonal basis (Grimes and Williams, 1990).

The loss of water from the leaves by transpiration is the driving force for the uptake of water from the soil. Other vine organs may also lose water, but at much slower rates than the leaves. The decrease in leaf water potential establishes a gradient in water potential between the leaf and the soil so that water flows into the vine's roots. Once in the root, water can move either through the apoplast, the water-filled space outside the cell membrane which includes the cell wall, or through the symplast, the protoplasm within the cell's membrane. Between the cortex and the vascular tissue of the root is the endodermis and a specialized cell layer called the Casparian strip. The latter consists of suberized tissue that restricts the flow of water through the cell-wall pathway.

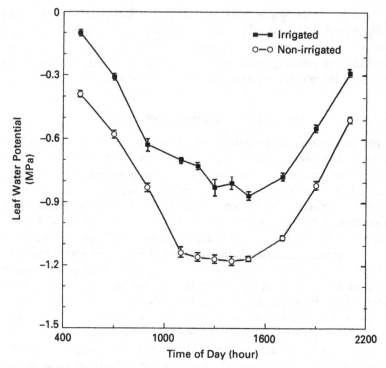

Fig. 4.10. Diurnal leaf water potential value of Thompson Seedless grapevines grown in the San Joaquin Valley of California. Vines were irrigated on 30 May 1989, or not irrigated at all since budbreak. Values represent the means of eight individual leaf replicates plus or minus one standard error

Water must pass through a cell membrane and cytoplasm before it can enter the xylem, a tissue that is considered part of the apoplast. The flow of water from the vascular tissue in the leaf to the substomatal cavity can be either apoplastic or symplastic. The flow of water through the vascular tissue is facilitated by the strong cohesive forces between water molecules and by capillary rise. The flow velocities of water in the xylem of woody species vary from 1.0 to 12 mm s^{-1} (Jones, 1983).

It has long been assumed that suberized roots are impermeable to water, but it is now known that those of woody species have hydraulic conductances of 40–70% of those of unsuberized roots when bathed in water (Bowen, 1985). This is supported by the observation that suberized roots of grapevines take up water early in the season (prior to anthesis) before there is appreciable new root growth.

Another major function of roots is the uptake of nutrients from the soil. There are two ways in which ions in the soil solution move towards the root surface: (i) mass flow in the water as it moves towards the root and (ii) diffusion. The concentration of mineral nutrient ions in the soil solution is generally quite low. Some ions, such as phosphorus, are usually bound to soil particles but others, such as nitrate, are mobile. In rapidly transpiring plants the uptake of nutrients into the root is by mass flow. In new roots the major absorption of ions is assumed to be through the region just behind the root tip. The absorption of ions in suberized roots, as with water, may not be as great as in those that are unsuberized. The absorption of ions by woody species such as *Vitis vinifera* is enhanced by the presence of vesicular – arbuscular mycorrhiza in the roots. It has been shown that mycorrhizal inoculation of grapevines increases the uptake of phosphorus and stimulates shoot growth (Possingham and Groot-Obbink, 1971; Gebbing *et al.*, 1977).

Once the nutrients enter the sap of the xylem in the root, they are carried to the shoot by mass flow in the transpiration stream. Most of these nutrients are in the form of inorganic ions. Nitrogen, sulfur and phosphorus may be carried in the form of organic compounds. In grapevines, nitrate is the predominate form of nitrogen taken up by the roots. It is carried to the leaves, where it is reduced in the chloroplasts for amino acid synthesis. Grape roots also have the ability to prevent the transport of salts, such as sodium, by retaining them in the root system (Jacoby, 1964).

The roots are also a source of phytohormones within the vine (Richards, 1983). Cytokinins are found in many plant organs, but the principal sites of synthesis are the roots. Roots are also a major source of gibberellins. Lastly, the production of abscisic acid in the roots when plants are subject to soil water deficits may be a means by which roots signal the leaves to close the stomata. The site of phytohormone synthesis in roots is at or close to the apex.

RELATION BETWEEN ROOT GROWTH AND GROWTH OF SHOOTS AND FRUITS

Flushes of root growth occur prior to anthesis and after harvest, but some elongation and production of new roots does occur at other times when the above-ground organs are in active growth. In addition, secondary growth and thickening occurs throughout the growing season.

Delaying the time of pruning of Thompson Seedless vines delays the spring flush of roots, and this indicates that the initiation of new roots is regulated by factors other than carbohydrates produced in the aerial portion of the vine. The factors concerned are probably hormonal but their identities are unknown. Indole-3-acetic acid (IAA), which is produced primarily in apices and expanding leaves, and which promotes root primordium formation, is a strong candidate.

The roots supply water and mineral nutrients needed for aerial growth, and the shoot supplies the necessary carbon substrates needed for root growth. The shoot : root ratio is used to express the quantitative relation between growth of the above-ground organs and growth of the roots. This ratio has been used extensively in annual plants, but its appropriateness in perennial plants such as grapevines is questionable because of the annual increase in dry mass of the permanent structures. The shoot : root ratio for field-grown Chenin blanc vines varied from 0.71 to 1.09 depending upon how the soil was prepared prior to planting (Saayman and van Huyssteen, 1980). Calculation of this ratio included dry masses of the trunk, cordons and shoots. Araujo (1988) found a shoot : root ratio of 5.2 in three-year-old Thompson Seedless vines grown in the field. In this research, calculations included dry masses of the shoots, fruiting canes and trunk. The ratio differed slightly between furrow-irrigated and drip-irrigated vines. The ratio of dry mass of the aerial portion of the vine to the root dry mass varies with the time of the year (Table 4.1).

Reserves of the grapevine

Carbohydrates (mostly starch) and mineral nutrients are stored as reserves within the grapevine. Reserve carbohydrates are utilized in the absence of recently formed photosynthate, for example, during the night period and after the leaves have fallen from the vine in autumn. Elements such as nitrogen and potassium are remobilized from storage organs and are translocated to growing points before there is appreciable mineral nutrient uptake in the spring or in circumstances of nutrient deficiency. Several researchers have attempted to characterize both carbohydrate (Winkler and Williams, 1945) and mineral reserves (Alexander, 1958) within vines, but have failed to provide quantitative estimates because they measured only concentrations. More recent studies have quantified the mineral elements required for growth of grapevines in pots (Conradie, 1980, 1981) and their remobilization from the vine's storage organs. However, it is seldom appropriate to extrapolate results with pot-grown vines to vines growing in a field situation.

CARBOHYDRATE RESERVES

The primary source of carbohydrates in vines for continued metabolic activity during the night period is the recently formed starch or soluble sugars stored in the leaves during the sunlight hours. The concentration

Table 4.2. *Soluble sugars and starch content (grams per vine) in organs of Chenin blanc grapevines*[a]

Organ	Mar. 19 sugars	Mar. 19 starch	May 27 sugars	May 27 starch	July 25 sugars	July 25 starch	Sept. 20 sugars	Sept. 20 starch
Roots	19	322	27	409	39	595	39	928
Trunk	22	184	15	173	21	302	19	404
Cordons	26	225	21	155	22	258	28	366
Stems	—	—	35	3	38	31	80	76
Leaves	—	—	44	<1	81	<1	84	7
Clusters	—	—	5	<1	589	16	2496	7

[a] Experimental details as in footnote to Table 4.1. Sugars refer to the combined total of glucose, fructose and sucrose. Sugars and starch were determined as described by Roper and Williams (1989).

of starch in the leaves varies on a diurnal basis; it is highest just before sunset and lowest at sunrise. The leaves also store carbohydrates if the trunk has been girdled.

The utilization of carbohydrate reserves that are stored in the permanent structures of perennial plants is a necessity early in the growing season. Unlike many deciduous fruit trees, which flower before the leaves emerge, grapevines do not flower until after considerable shoot growth has occurred. This means that their requirements for reserves to support early-season growth may be less than those of some other perennial crops because the newly emergent shoots quickly become self-sufficient for carbohydrates. From budbreak until two weeks after anthesis, approximately 95 g of non-structural carbohydrates (NSC) were lost from the trunk and cordons of Chenin blanc grapevines (Table 4.2). These reserves may have been mobilized to the new shoots until they became self-sufficient, or to the root system: an increase in the non-structural carbohydrate content in the roots (95 g per vine) was recorded during this period. The amount of non-structural carbohydrates continued to increase in the roots, cordons and trunk until the time of harvest. The amount of NSC in the roots was 2.7 times greater at harvest than at budbreak whereas the trunk and cordons had, respectively, 1.8 and 1.5 times more NSC. The smaller amount of NSC at budbreak indicates that the reserve carbohydrates in those structures were utilized for maintenance respiration throughout the dormant period. The continued increase in sugars after harvest may be of special importance in regions with very low winter temperatures, because the elevated solute content of canes, spurs and trunk may protect the vines from freezing damage.

Table 4.3. *Nitrogen and potassium content (grams per vine) of Chenin blanc grapevines during the growing season[a]*

Organ	Mar. 19 N	Mar. 19 K	May 27 N	May 27 K	July 25 N	July 25 K	Sept. 20 N	Sept. 20 K
Roots	90	26	80	21	63	22	68	26
Trunk	15	26	15	37	21	26	21	33
Cordons	23	22	25	36	27	26	28	32
Stems	—	—	9	24	35	31	32	44
Leaves	—	—	44	20	37	36	32	29
Clusters	—	—	11	21	23	48	37	88

[a] Experimental details as in footnote to Table 4.1. Nitrogen and potassium were determined as described by Williams (1987b) and Williams *et al.* (1987), respectively.

MINERAL NUTRIENT RESERVES

The reserves of mineral nutrients, especially nitrogen, are important in the overall growth of the grapevine. Arginine is the main form of storage N in the grapevine (Kliewer, 1967). It is assumed that a large percentage of the N required for new shoot growth is remobilized from existing N reserves in the permanent structures of the vine, predominantly from the roots. The loss of N from the root system accounted for approximately 40% of the N needed by new shoots in young pot-grown vines (Conradie, 1980). In field-grown vines, 14–26% of the N required for new shoot growth was remobilized from permanent organs other than the roots (Araujo and Williams, 1988). For mature Chenin blanc vines, only 10 g N vine^{-1} was lost from the permanent structures of the vine between budbreak and shortly after anthesis (Table 4.3). This represents less than 15% of the N required by the new shoots and fruit during that period. In a different study on Thompson Seedless grapevines, 15 g N vine^{-1} were remobilized from the roots to the shoots between budbreak and bloom; and this represented 70% of the N requirements of the shoots. The amount of N in the root system continues to decrease until the last week in July and then starts to increase toward harvest. The pattern in Table 4.3 demonstrates that roots are able to supply the rest of the vine with N taken up from the soil even early in the growing season. There is also utilization of vine N reserves later in the season. Using ^{15}N-labeled fertilizer, it was determined that N accumulated in the fruit was derived primarily from N stored from the roots and mature wood (Conradie, 1981). The amount of N remobilized from the roots, trunk and other permanent structures is dependent upon vine age, time of the year and growth conditions.

The distribution of potassium within the vine differs from that of N. Absolute amounts of K in the roots are less than those of N, whereas in the trunk and cordons K generally is greater than N (Table 4.3). Only a small amount of K is remobilized from the roots, but none of the K in the trunk and cordons is remobilized to other organs. Fruit is the major sink for K after berry growth commences. Several studies have shown that remobilization of K from leaves to the fruit occurs if the vine's canopy is extremely dense. Other studies have shown that there may be some redistribution of K from the main axis of the shoot to the clusters. However, the majority of K found in the fruits is taken up from the soil. Little is known of the storage and utilization of other mineral nutrients within the vine.

NUTRITIONAL ASPECTS OF LEAF SENESCENCE AND ABSCISSION

According to some authorities, the processes of senescence of leaves and abscission involve, and are preceded by, the recovery of mineral nutrients and carbohydrates by the permanent structures (Grigal *et al.*, 1976; Oland, 1963). Shedding of leaves occurs when the abscission zones in the leaf and petiole are activated. Little is known of the hormonal control of leaf senescence and abscission in grapevines, but their regulation is likely to be similar to that in other deciduous perennials. In perennial crops, it is often assumed that a considerable fraction of the mineral content of senescing leaves is translocated back to the permanent structures. In grapevines the leaves contain a large proportion of the total N content, but little of it is remobilized to the trunk or roots (Williams, 1987*b*). Less than 5 g N vine^{-1} was translocated back into the vine, whereas approximately 30 g N vine^{-1} (33 kg N ha^{-1}) remained in the leaves that fell to the ground. In addition, the equivalent of 7 kg K ha^{-1} was present in the fallen leaves (Williams *et al.*, 1987). It appears that remobilization of mineral nutrients from the leaves of grapevines as they senesce is of little significance in terms of maintaining internal levels, but the mineral nutrients of leaves and prunings, once decomposed in the soil, do become available to the vine in the following years. At present, there is little information on the extent of mineral nutrient recycling in the vineyard.

Literature cited

Alexander, D.M. 1958. Seasonal fluctuations in the nitrogen content of the Sultana vine. *Aust. J. Agric. Res.* **8**: 162–78.
Amerine, M.A. and Winkler, A.J. 1944. Composition and quality of musts and wines of California grapes. *Hilgardia* **15**: 493–675.

Amthor, J.S. 1989. *Respiration and crop productivity.* New York: Springer-Verlag.

Antcliff, A.J. and May, P. 1961. Dormancy and bud burst in Sultana vines. *Vitis* **3**: 1–14.

Araujo, F.J. 1988. *The response of three year-old Thompson Seedless grapevines to drip and furrow irrigation in the San Joaquin Valley.* Masters thesis, University of California, Davis.

Araujo, F.J. and Williams, L.E. 1988. Dry matter and nitrogen partitioning and root growth of young 'Thompson Seedless' grapevines grown in the field. *Vitis* **27**: 21–32.

Archer, E. and Strauss, H.C. 1985. Effect of plant density on root distribution of three-year-old grafted 99 Richter grapevines. *S. Afr. J. Enol. Vitic.* **6**: 25–30.

Atkinson, D. 1980. The distribution and effectiveness of roots of tree crops. *Hort. Reviews* **2**: 424–490.

Baggiolini, M. 1952. *Les stades repères dans le developpement annuel de la vigne et leur utilisation pratique.* Stn. Fed. Essais Agric. (Lausanne), Publ. 12 (MC). 3 pp.

Baldwin, J.G. 1966. The effect of some cultural practices on nitrogen and fruitfulness in the Sultana vine. *Am. J. Enol. Vitic.* **17**: 58–62.

Beck, E. and Ziegler, P. 1989. Biosynthesis and degradation of starch in higher plants. *Ann. Rev. Plant Physiol. Plant Molec. Biol.* **40**: 95–117.

Bernstein, L., Ehlig, C.F. and Clark, R.A. 1969. Effect of grape rootstocks on chloride accumulation in leaves. *J. Amer. Soc. Hort. Sci.* **94**: 584–90.

Bowen, G.D. 1985. Roots as a component of tree productivity. In *Attributes of trees as crop plants.* (ed. M.G.R. Cannell and J.E. Jackson), pp. 303–15. Titus Wilson & Son, Kendal, Cumbria.

Buttrose, M.S. and Hale, C.R. 1971. Effects of temperature on accumulation of starch or lipid in chloroplasts of grapevine. *Planta* **101**: 166–70.

Chatterton, J.J. and Silvius, J.E. 1979. Photosynthate partitioning into starch in soybean leaves. I. Effect of photoperiod *versus* photoperiod duration. *Plant Physiol.* **64**: 749–53.

Conradie, W.J. 1980. Seasonal uptake of nutrients by Chenin blanc in sand culture. I. Nitrogen. *S. Afr. J. Enol. Vitic.* **1**: 59–65.

Conradie, W.J. 1981. Seasonal uptake of nutrients by Chenin blanc in sand culture. II. Phosphorus, potassium, calcium and magnesium. *S. Afr. J. Enol. Vitic.* **2**: 7–13.

Coombe, B.G. 1960. Relationship of growth and development to changes in sugars, auxins, and gibberellins in fruit of seeded and seedless varieties of *Vitis vinifera.* *Plant Physiol.* **35**: 241–50.

De La Harpe, A.C. and Vissor, J.H. 1985. Growth characteristics of *Vitis vinifera* L. cv. Cape Riesling. *S. Afr. J. Enol. Vitic.* **6**: 1–6.

Downton, W.J.S. 1985. Growth and mineral composition of the Sultana grapevine as influenced by salinity and rootstock. *Aust. J. Agric. Res.* **36**: 425–34.

Downton, W.J.S., Grant, W.J.R. and Loveys, B.R. 1987. Diurnal changes in the photosynthesis of field-grown grape vines. *New Phytol.* **105**: 71–80.

During, H. 1978. Studies on the environmentally controlled stomatal transpiration in grape vines. II. Effects of girdling and temperatures. *Vitis* **17**: 1–9.

Ehlig, C.F. 1960. Effect of salinity on four varieties of table grapes grown in sand culture. *Proc. Amer. Soc. Hort. Sci.* **76**: 323–35.

Eichhorn, K.W. and Lorenz, D.H. 1977. Phänologische entwicklungsstadien der rebe. *Nachichtenbl. Dtsch. Pflanzenschutzdienstes (Braunschweig)* **29**: 119–120.

Emmerson, J.G. and Powell, L.E. 1978. Endogenous abscisic acid in relation to rest and bud burst in three *Vitis* species. *J. Amer. Soc. Hort. Sci.* **103**: 677–90.

Freeman, B.M. and Smart, R.E. 1976. Research note: A root observation laboratory for studies with grapevines. *Am. J. Enol. Vitic.* **27**: 36–9.

Gebbing, H., Schwab, A. and Alldweldt, G. 1977. Mycorrhiza of vines. *Vitis* **16**: 279–85.

Giaquinta, R. 1983. Phloem loading of sucrose. *Ann. Rev. Plant Physiol.* **34**: 347–87.

Grigal, D.F., Ohmann, L.F. and Brander, R.B. 1976. Seasonal dynamics of tall shrubs in northeastern Minnesota: biomass and nutrient element changes. *For. Sci.* **22**: 195–208.

Grimes, D.W. and Williams, L.E. 1990. Irrigation effects on plant water relations and productivity of Thompson Seedless grapevines. *Crop Sci.* **30**: 255–60.

Gutierrez, A.P., Williams, D.W. and Kido, H. 1985. A model of grape growth and development: The mathematical structure and biological considerations. *Crop Sci.* **25**: 721–8.

Hale, C.R. and Weaver, R.J. 1963. The effect of developmental stage on direction and translocation of photosynthate in *Vitis vinifera*. *Hilgardia* **33**: 89–131.

Harrell, D.C. and Williams, L.E. 1987. Net CO_2 assimilation rate of grapevine leaves in response to trunk girdling and gibberellic acid application. *Plant Physiol.* **83**: 457–9.

Hiroyasu, T. 1961. Nutritional and physiological studies on the grapevine. *J. Jap. Soc. Hort. Sci.* **30**: 111–16.

Hofacker, W. 1978. Investigation on the photosynthesis of vines. Influence of defoliation, topping, girdling and removal of grapes. *Vitis* **11**: 10–22.

Huber, S.C. 1986. Fructose 2, 6-biophosphate as a regulatory metabolite in plants. *Ann. Rev. Plant Physiol.* **37**: 233–46.

Huglin, P. 1958. Recherches sur les bourgeons de la vigne: initiation florale et développement vegetatif. *Ann. Amélior. Plantes* **8**: 113–272.

Husic, D.W., Husic, H.D. and Tolbert, N.E. 1987. The oxidative photosynthetic carbon cycle or C_2 cycle. In *CRC Critical Reviews in Plant Sciences*, vol. 5, issue 1, pp. 45–100. CRC Press, Orlando.

Jacoby, B. 1964. Function of the root and stems in sodium retention. *Plant Physiol.* **39**: 445–9.

Jones, H.G. 1983. *Plants and microclimate.* Cambridge University Press. 323 pp.

Kliewer, W.M. 1967. Annual cyclic changes in the concentration of free amino acids in grapevines. *Am. J. Enol. Vitic.* **18**: 126–37.

Kramer, P.J. and Kozlowski, T.T. 1979. *Physiology of woody plants.* Academic Press, Orlando, Florida. 811 pp.

Kriedemann, P.E. 1968. Photosynthesis in vine leaves as a function of light intensity, temperature, and leaf age. *Vitis* **7**: 213–20.

Kriedemann, P.E., Loveys, B.R. and Downton, W.J.S. 1975. Internal control of stomatal physiology and photosynthesis. II. Photosynthetic responses to phaseic acid. *Aust. J. Plant Physiol.* **2**: 553–67.

Loveys, B.R. and During, H. 1984. Diurnal changes in water relations and abscisic acid in field-grown *Vitis vinifera* cultivars. II. Abscisic acid changes under semi-arid conditions. *New Phytologist* **97**: 37–47.

McIntyre, G.N., Lider, L.A. and Ferrari, N.L. 1982. The chronological classification of grapevine phenology. *Am. J. Enol. Vitic.* **33**: 80–5.

McIntyre, G.N., Kliewer, W.M. and Lider, L.A. 1987. Some limitations of the degree day system as used in viticulture in California. *Am. J. Enol. Vitic.* **38**: 128–32.

McKenry, M.V. 1984. Grape root phenology relative to control of parasitic nematodes. *Am. J. Enol. Vitic.* **35**: 206–11.

Moncur, M.W., Rattigan, K., Mackenzie, D.H. and McIntyre, G.N. 1989. Base temperature for budbreak and leaf appearance of grapevines. *Am. J. Enol. Vitic.* **40**: 21–6.

Münch, E. 1930. *Die Stoffbewegungen in der Pflanze*. Gustav Fischer, Jéna., 234 pp.

Nir, G., Klein, I., Lavee, S., Spieler, G. and Barak, U. 1988. Improving grapevine budbreak and yields by evaporative cooling. *J. Amer. Soc. Hort. Sci.* **113**: 512–17.

Oland, K 1963. Changes in the content of dry matter and major nutrient elements of apple foliage during senescence and abscission. *Physiol. Plant.* **26**: 682–94.

Osmond, C.B. 1981. Photorespiration and photoinhibition: Some implications for the energetics of photosynthesis. *Biochim. Biophys. Acta* **639**: 11–98.

Patrick, J.W. 1990. Sieve element unloading: cellular pathway, mechanism and control. *Physiol. Plant.* **78**: 298–308.

Possingham, J.V. and Groot-Obbink, J. 1971. Endotrophic mycorrhiza and the nutrition of grapevines. *Vitis* **10**: 120–130

Pouget, R. and Rives, M. 1958. Action de la rindite sur la dormance de la vigne (*Vitis vinifera* L.). *C.R. Acad. Sci. Paris* **246**: 3664–6.

Powles, S.B. 1984. Photoinhibition of photosynthesis induced by visible light. *Ann. Rev. Plant Physiol.* **35**: 15–44.

Pratt, C. 1971. Reproductive anatomy in cultivated grapes – A review. *Amer. J. Enol. Vitic.* **22**: 295–8.

Raschke, K. 1979. Movements of stomata. In *Encyclopedia of plant physiology* (new series), vol. 7, (ed. W. Haupt and M.E.F. Feinleib), pp. 383–441. Springer-Verlag, Berlin.

Richards, P. 1983. The grape root system. *Hort. Reviews* **5**: 127–68.

Roper, T.R. and Williams, L.E. 1989. Net CO_2 assimilation and carbohydrate partitioning of grapevine leaves in response to trunk girdling and gibberellic acid application. *Plant Physiol.* **89**: 1136–40.

Saayman, D., and L. Van Huyssteen. 1980. Soil preparation studies: I. The effect of depth and method of soil preparation and of organic material on the performance of *Vitis vinifera* (var. Chenin Blanc) on Hutton/Sterkspruit Soil. *S. Afr. J. Enol. Vitic.* **1**: 107–21.

Seguin, M.G. 1972. Répartition dans l'espace du systeme radiculaire de la vigne. *C.R. Acad. Sci. Paris* D: 2178–80.

Sharkey, T.D. 1988. Estimating the rate of photorespiration in leaves. *Physiol. Plant.* **73**: 147–53.

Shulman, Y., Nir, G., Fanberstein, L. and Lavee, S. 1983. The effect of cyanamide on the release from dormancy of grapevine buds. *Sci. Hort.* **19**: 99–104.

Skinner, P. and Matthews, M.A. 1990. A novel interaction of magnesium translocation with a supply of phosphorus to roots of grapevines, *Vitis vinifera* L. *Plant Cell Environ.* **13**: 821–6.

Smart, R.E., Robinson, J.B., Due, G.R. and Brien, C.J. 1985. Canopy microclimate modification for the cultivar Shiraz. I. Definition of canopy microclimate. *Vitis* **24**: 17–31.

Southey, J.M. and Archer, E. 1988. The effect of rootstock cultivar on grapevine root distribution and density. In *The grapevine root and its environment* (J.L. van Zyl (Comp.) Tech. Comm. 215), pp. 57–73. Dept. Agric. Water Supply, Pretoria.

Swanepoel, J.J. and Southey, J.M. 1989. The influence of rootstock on the rooting pattern of the grapevine. *S. Afr. J. Enol. Vitic.* **10**: 23–8.

Turgeon, R. 1989. The sink-source transition in leaves. *Ann. Rev. Plant Physiol. Plant Molec. Biol.* **40**: 119–38.

van Zyl, J.L. 1984. Response of Colombard grapevines to irrigation as regards quality aspects and growth. *S. Afr. J. Enol. Vitic.* **5**: 19–28.

Weaver, R.J. 1959. Prolonging dormancy in *Vitis vinifera* with gibberellin. *Nature* **183**: 1198–9.

Williams, D.W., Williams, L.E., Barnett, W.W., Kelley, K.M. and McKenry, M.V. 1985*a*. Validation of a model for the growth and development of the Thompson Seedless grapevine. I. Vegetative growth and fruit yield. *Amer. J. Enol. Vitic.* **36**: 275–82.

Williams, D.W., Andris, H.L., Beede, R.H., Luvisi, D.A. Norton, M.V.K. and Williams, L.E. 1985*b*. Validation of a model for the growth and development of the Thompson Seedless grapevine. II. Phenology. *Amer. J. Enol. Vitic.* **36**: 283–9.

Williams, L.E. 1987*a*. Growth of 'Thompson Seedless' grapevines: I. Leaf area development and dry weight distribution. *J. Amer. Soc. Hort. Sci.* **112**: 325–30.

Williams, L.E. 1987*b*. Growth of 'Thompson Seedless' grapevines: II. Nitrogen distribution. *J. Amer. Soc. Hort. Sci.* **112**: 330–3.

Williams, L.E. 1987*c*. The effect of cyanamide on budbreak and vine development of Thompson Seedless grapevines in the San Joaquin Valley of California. *Vitis* **2**: 107–13.

Williams, L.E. and Biscay, P.J. 1991. Partitioning of dry weight, nitrogen and potassium in Cabernet Sauvignon grapevines from anthesis until harvest. *Am. J. Enol. Vitic.* **42**: 113–17.

Williams, L.E., Biscay, P.J. and Smith, R.J. 1987. Effect of interior canopy defoliation on berry composition and potassium distribution of Thompson Seedless grapevines. *Amer. J. Enol. Vitic.* **38**: 287–92.

Williams, L.E. and Grimes, D.W. 1987. Modelling vine growth – development of a data set for a water balance subroutine. In *Proceedings of the 6th Australian Wine Industrial and Technical Conference, Adelaide, Australia, 14–17 July 1986*, (ed. T. Lee), pp. 169–74. Adelaide: Australian Industrial Publishers.

Williams, L.E. and Matthews, M.A. 1990. Grapevine. In *Irrigation of Agricultural Crops* (Agronomy Monographs no. 30) (ed. B.J. Stewart and D.R. Nielsen), pp. 1019–55. ASA-CSSA-SSSA, Madison, Wisconsin.

Williams, L.E., Neja, R.A., Meyer, J.L., Yates L.A. and Walker, E.L. 1991. Postharvest irrigation influences budbreak of 'Perlette' grapevines. *Hort Sci.* **26**: 1081.

Williams, L.E. and Smith, R.J. 1985. Net CO_2 assimilation rate and nitrogen content of grape leaves subsequent to fruit harvest. *J. Amer. Soc. Hor. Sci.* **110**: 846–50.

Williams, L.E. and Smith, R.J. 1991. Partitioning of dry weight, nitrogen and potassium and root distribution of Cabernet Sauvignon grapevines grafted on three different rootstocks. *Amer. J. Enol. Vitic.* **42**: 118–22.

Winkler, A J. and Williams, W.O. 1945. Starch and sugars of *Vitis vinifera*. *Plant Physiol.* **20**: 412–32.

Yang, Y. and Hori. Y. 1979. Studies on retranslocation of accumulated assimilates in 'Delaware' grapevines. I. Retranslocation after ^{14}C feeding in Summer and Autumn. *Tohoku J. Agric. Res.* **30**: 43–56.

5
Developmental physiology: flowering and fruiting

Regulation of flowering in the grapevine

The complex morphology of the grapevine shoot system, and the origin of inflorescences, were described in detail in Chapter 3. To recapitulate, flowering in the mature grapevine is normally a three-step process. The first step is the formation of *Anlagen or* 'uncommitted primordia' by the apices of specialized lateral buds (latent buds) on shoots of the current season. Next, the Anlagen develop either as inflorescence primordia or as tendril primordia, and shortly thereafter the latent buds enter into dormancy. In some circumstances Anlagen produce shoots instead of tendrils or inflorescences. Finally, the formation of flowers from the inflorescence primordia occurs at the time of bud burst in the next season. This chapter is concerned with the effects of phytohormones and growth regulators on the constituent steps in the flowering process and with the effects of environment and nutrition on flowering.

HORMONAL ASPECTS OF FLOWERING

REGULATION OF ANLAGEN AND TENDRIL FORMATION

The first step in the development of the Anlage is the formation of a bract (Fig. 5.1). Next, the apex subtended by the bract divides to form two branches or arms. This is a crucial stage in the reproductive development of the grapevine because two-branched Anlagen have the potential to produce inflorescence primordia, tendril primordia or shoot primordia (buds). Shoot formation from Anlagen is rare, and its control is not well understood. In the case of flowering, however, it is evident that control can be exercised at two levels. The first is a coarse control and involves the formation of Anlagen. The second is a finer level of control and involves the switching of the two-branched Anlage into either the inflorescence or the tendril pathway.

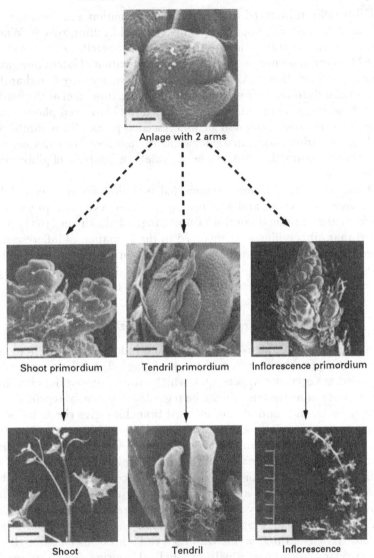

Fig. 5.1. Pathways of Anlage development. Anlage with two arms (bar = 64 μm); shoot primordium (bar = 230 μm); tendril primordium (bar = 105 μm); inflorescence primordium (bar = 400 μm); shoot (bar = 53 mm); tendril (bar = 800 μm); inflorescence (bar = 32 mm). Dotted lines refer to potential to switch from one pathway to another under influence of phytohormones. From Mullins (1980). Reproduced with permission

Gibberellin is involved in both Anlagen formation and the determination of Anlagen development (Srinivasan and Mullins, 1980a). When container-grown grapevines are treated with gibberellic acid (GA$_3$, 3–30 μM) there is premature sprouting and elongation of latent buds and precocious formation of Anlagen, i.e. the first Anlagen are found at the second and third nodes from the base of the stem instead of at the fourth or fifth nodes as is normal. Anlagen formed in GA-treated plants grow only into tendrils; formation of inflorescence primordia is inhibited. Anlage formation and tendril elongation are suppressed by chlormequat (2-chloroethyltrimethylammonium chloride), an inhibitor of gibberellin biosynthesis.

Anlagen and tendrils may be regarded as the inflorescence axes of the grapevine (Srinivasan and Mullins, 1979). After reviewing 40 years of work on the hormonal control of flowering, Chailakhyan (1977) proposed that gibberellins were involved in the formation of inflorescence axes in plants. A requirement for gibberellin for the formation of inflorescence axes in grapes is strongly suggested by the inhibition of Anlage formation and the suppression of tendril growth by chlormequat.

DIFFERENTIATION OF INFLORESCENCES

A characteristic of *Vitis* is that tendrils and inflorescences are homologous organs, both arising from the same Anlage. Tendrils are generally regarded as vegetative appendages, which provide support for climbing plants, but grapevine tendrils can be regarded as potential reproductive organs. Anlagen that undergo repeated branching give rise to inflorescence primordia, but those that produce only two or three branches give rise to tendrils (Srinivasan and Mullins, 1976). Accordingly, the regulation of inflorescence formation in the grapevine hinges on the control of branching of Anlagen (or of tendrils); this process has been studied by growing isolated apices and tendrils *in vitro* with phytohormones and growth regulators and by application of growth substances to the apices and tendrils of intact grapevines (Srinivasan and Mullins, 1978, 1979, 1980a). When cultured *in vitro* with the cytokinins benzyladenine (BA), 6-(benzylamino)-9-(2-tetrahydropyranyl)-9H purine (PBA) or zeatin riboside, isolated tendrils were induced to undergo repeated branching and they grew into inflorescence and inflorescence-like structures. There was normal development of calyx and calyptra, but the flowers lacked functional ovules and anthers. With intact plants, repeated applications of PBA (50–200 μM) to shoot tips led to the transformation of newly formed tendrils into inflorescences, which subsequently flowered and set fruit containing viable seeds.

Unlike the situation with Anlage formation, inflorescence formation from Anlagen and from tendrils is favored by treatment with chlormequat. Cytokinin induces the formation of inflorescences from Anlagen and from young tendrils, and there are a few reports that chlormequat increases endogenous cytokinin levels in the ascending sap of the grapevine (Skene, 1970; Lilov and Andanova, 1976). Generally, chlormequat-induced alterations in the pattern of growth and development of grapevines are the opposite of those induced by GA, but some of the effects of chlormequat, such as the production of dark green leaves, formation of secondary inflorescences (Coombe, 1967) and increased fruit set (Coombe, 1970), seem to mimic the effect of cytokinins. It is possible, therefore, that chlormequat may have a dual role in regulating inflorescence formation in the grapevine: (i) inhibition of gibberellin synthesis; and (ii) enhancement of cytokinin production.

INDUCTION OF PRECOCIOUS FLOWERING IN GRAPEVINE SEEDLINGS

Tendrils of grapevine seedlings are converted into inflorescences when treated with cytokinins, but chlormequat does not induce flowering when applied alone. However, when chlormequat sprays (250–500 µM) are followed by, or combined with, treatment of tendrils with PBA (500–1000 µM) there is prolific flower formation. These observations are consistent with the view that variation in the endogenous levels of gibberellin and cytokinin is the factor that determines whether the Anlage will grow into a tendril or an inflorescence.

The ease with which tendrils may be converted into inflorescences, and the size of the bunches of grapes produced, varies with genotype and the age of the seedling when growth regulators are applied. Seedlings of Muscat Hamburg and Gloryvine were found to be more favorable materials than seedlings of Cabernet Sauvignon for induction of flowering by growth regulators. The size of inflorescence increases with increasing age of the seedling. The induction of precocious flowering in grapevine seedlings by growth regulators is clearly of great interest in grapevine breeding as a means of speeding the generations (Srinivasan and Mullins, 1981a,b).

EFFECTS OF CYTOKININ ON DIFFERENTIATION OF FLOWERS

When dormant latent buds are activated in spring, the inflorescence primordia that were formed in the previous summer undergo rapid development to form the flower primordia. Flower formation is a

cytokinin-controlled process. The xylem sap (bleeding sap) of the grapevine contains high cytokinin activity during bud burst and flowering
(Nitsch and Nitsch, 1965; Skene and Kerridge, 1967). Moreover, there
is strong evidence that cytokinin produced by roots is involved in the
regulation of flower differentiation. The inflorescence primordia of
hardwood cuttings shrivel and die if the emergence of the inflorescence
precedes the formation of roots, but normal inflorescences are formed
when cuttings are propagated by a special technique, which ensures
adventitious root formation in advance of bud burst (Mullins, 1966).
Rootless cuttings require exogenous cytokinins (BA or PBA) for normal
differentiation of inflorescences (Mullins, 1967, 1968). Other effects of
exogenous cytokinin on the reproductive development of grapevines include the induction of pistil development in male genotypes (Negi and
Olmo, 1966) and the promotion of fruit set (Weaver *et al.*, 1965).

SUMMARY: THE HORMONAL CONTROL OF FLOWERING

The concept of a single trigger for flowering is inappropriate in the
grapevine because inflorescence formation is regulated at two levels:
formation of Anlagen and differentiation of Anlagen. The theory that
the floral stimulus involves two complementary stimuli, as suggested by
Carr (1968), Chailakhyan (1977) and Evans (1971), is a more attractive
hypothesis. Zeevaart (1976) proposed that the requirement for a specific
balance of hormones for flower formation is readily applicable to woody
perennials. Evidence is accumulating in the grapevine that gibberellin
and cytokinin are the principal regulators of flowering.

Chailakhyan (1977) has suggested that gibberellins are involved in
the formation and growth of floral stems or inflorescence axes. The responses of grapes to exogenous GA and chlormequat are consistent with
this view. Gibberellin is necessary for the formation of inflorescence axes
(initiation of Anlagen) and for the growth of inflorescence axes (two-
branched stage of the Anlagen).

Gibberellins are inhibitors of flowering in many fruit species, but the
role of GA in flowering in grapevines varies with the stage of development of the latent bud. At an early stage, GA is a promoter of flowering
because Anlagen formation is a GA-requiring process. Later, GA acts as
an inhibitor of flowering because it directs the Anlagen to form tendrils.

Cytokinins are implicated in the control of many aspects of reproduction in the grapevine. The mechanism by which cytokinins exert these
effects is unknown but may be associated with the partitioning of assimilates. It has been demonstrated in many plants, including *Vitis vinifera*,
that cytokinins are strong mobilizers of assimilates to the site of applica-

tion. In another perennial, *Bougainvillea*, Tse *et al.* (1974) showed that PBA-induced accumulation of ^{14}C-assimilates was followed by inflorescence formation, and several other authors have suggested that redistribution of metabolites is involved in the regulation of flowering (Sachs and Hackett, 1976; Sachs, 1977).

ENVIRONMENTAL FACTORS IN FLOWERING

TEMPERATURE

There are many reports of a requirement for high temperatures for inflorescence primordium formation in grapes (Srinivasan and Mullins, 1981*a*). There is a positive relation between temperature from the middle of June to the middle of July (northern hemisphere) and the number of inflorescences appearing on the shoot in the following season (Alleweldt, 1963). Specifically, high temperatures during stages 5–7 of latent bud development are closely correlated with subsequent fruitfulness of latent buds (Baldwin, 1964).

It is difficult to demonstrate a specific effect of temperature on flower formation under vineyard conditions. To circumvent this difficulty, Buttrose (1969*a,b,c*) used grapevines grown in small containers in controlled environments. After three months of growth, the latent buds were dissected under a stereomicroscope, and the size and mass of inflorescence primordia were measured. With the relatively fruitful cultivar Muscat of Alexandria, the number of inflorescence primordia recognizable at three months varied from zero at 20 °C to a maximum of 1.6 in vines grown at temperatures close to 35 °C. A pulse of only four hours per day (or night) of high temperature (30 °C) was sufficient to induce the maximum number of inflorescence primordia (stage 7). The critical period for susceptibility to the high-temperature response is the three weeks before the formation of Anlagen by the apices of latent buds.

Substantial differences have been found in temperature requirements for inflorescence primordium formation among cultivars of different geographic origin. Riesling and Shiraz initiate inflorescences with temperatures as low as 20 °C, but Muscat of Alexandria requires a temperature of 25 °C (Buttrose, 1970*a,b*). Cabernet Sauvignon requires a lower temperature summation for flowering than Bulgar or Rkatsiteli (Braikov, 1975). A high temperature pulse is essential for the initiation of the second and third inflorescence in many cultivars, including cool-climate cultivars. Thompson Seedless and Ohanez are less fruitful than most other cultivars and are more responsive to changes in temperature. American cultivars (interspecific hybrids) such as Delaware produce inflorescences at lower temperatures (21–22 °C) than do *V. vinifera* cultivars (27–28 °C for Muscat of Alexandria).

LIGHT INTENSITY

Effects of light intensity on the fruitfulness of grapevine buds are independent of temperature (Buttrose, 1970*b*). The effect of light intensity on inflorescence formation has been studied in the vineyard in relation to the hours of sunshine (Baldwin, 1964), or to shading treatments (May and Antcliff, 1963). Shading reduces fruitfulness. A mean of 10 h sunshine per day during inflorescence formation is needed for an acceptable level of fertility in Thompson Seedless vines. Shading for four weeks during late spring reduces the fruitfulness of latent buds to a greater extent than shading treatments applied earlier or later during the season. In New Zealand, shoots covered with hessian shades (26% full sunlight) during December to May (summer and autumn, southern hemisphere) produced fewer bunches than unshaded shoots (Hopping, 1975). However, growth cabinet studies showed that illumination equivalent to one quarter of full sunlight (39 klx) is sufficient to obtain maximum fruitfulness in container-grown plants of Muscat of Alexandria (Buttrose, 1974*a*). In vines grown in controlled environments, the number and size of inflorescence primordia increases with increases in light intensity (Buttrose, 1968).

Vertically trained shoots are more fruitful than horizontally trained shoots (May, 1966). Direct exposure of latent buds to high-intensity light improves the fruitfulness of buds, but no effects of the quality of light on inflorescence formation have been reported.

It is often observed that buds situated inside the canopy of field-grown vines are less fruitful than those at the exterior where buds are more strongly illuminated (May *et al.*, 1976). The use of trellises and split canopies, as in the Geneva Double Curtain system of training, gives improved fruitfulness of buds and an overall increase in productivity of 50–90% (Shaulis *et al.*, 1966).

As with the effect of temperature, responses of vines to differing light intensities vary with the cultivar. Thompson Seedless, Ohanez, and Shiraz were fruitful only at light intensities higher than 19.5 klx but Muscat of Alexandria and Rhine riesling were fruitful with an illumination of 19.5 klx (Buttrose, 1970*b*).

PHOTOPERIOD

Photoperiod does not affect inflorescence induction in grapes, but there is evidence in some cultivars that the numbers of inflorescence primordia per bud are greater under long days than with short days (Buttrose, 1974a). The fruitfulness of vines grown in growth cabinets was promoted by increasing the time of exposure to high-intensity light (3600 ft candles (38.8 klx)). These results could not be explained by the increase in quantity of light for use in photosynthesis. If the high light intensity was supplied for more than 12 h per day, the formation of inflorescence primordia appeared to depend on the hours of illumination rather than on total incident energy. In contrast, the accumulation of dry matter was related to total incident light energy and not to the number of hours of illumination (Buttrose, 1968). The mechanism leading to inflorescence primordium formation does not seem to be closely related to the mechanism of dry mass accumulation (i.e. photosynthesis) despite its requirements for high-energy light.

American species, including *Vitis labrusca*, are more sensitive to day length than *Vitis vinifera* L. (Kobayashi *et al.*, 1965, 1966; Sigiura *et al.*, 1975). Delaware vines (*Vitis* × *labruscana*) grown in long days formed nearly three times as many inflorescences as those grown in short days, irrespective of the temperature regime.

To sum up the influence of temperature and light on fruitfulness, Buttrose (1974a) concluded that temperature is a dominant factor for inflorescence primordia formation, but according to Rives (1972), light intensity is the limiting factor. It is probable that a combination of exposure to high temperature and high light intensity is necessary for maximum fruitfulness of latent buds.

WATER STRESS

Persistent water stress depresses the fruitfulness of latent buds; this explains why rain-fed vines usually bear fewer fruitful buds than irrigated vines (Huglin, 1960). Soil moisture is one of the chief factors influencing inflorescence development in grapes (Alleweldt and Hofacker, 1975); studies with vines grown in controlled environments have shown that the number and size of inflorescence primordia is reduced by water stress (Buttrose, 1974b). However, there are also reports that water stress increases the fruitfulness of buds (Smart *et al.*, 1974). May (1965) has suggested that the reduced foliage density of water-stressed vines improves the illumination within the canopy and results in improved fertility of basal buds, a factor that leads to increased fruitfulness of the vine as a whole.

Shoot growth is sensitive to water stress: there is a reduction in both bud fruitfulness and dry mass of shoots (Buttrose, 1974b). This suggests that water stress may affect fruitfulness indirectly by reducing photosynthesis (Loveys and Kriedemann, 1973). Moreover, water stress causes a decrease in cytokinin in the xylem sap (Livne and Vaadia, 1972) and an increase in the abscisic acid levels in leaves and stems (Düring and Alleweldt, 1973).

Information on the relation between water stress and fruitfulness of latent buds is limited because of the difficulty of separating the specific effects of water stress from those of temperature and light intensity.

MINERAL NUTRITION

Most studies on the mineral nutrition of grapes have been concerned with berry development and wine quality, and there are few reports on the effects of mineral nutrition on flower formation. An adequate supply of nitrogen is necessary for inflorescence primordium formation and for the differentiation of flowers (Alleweldt, 1964). Size of inflorescence primordia is generally little affected by N nutrition (Srinivasan et al., 1972), but an increase in the number of inflorescence primordia following N application is found when the initial N status of the vine is low (Baldwin, 1966). Under special circumstances, application of nitrogen can result in a reduction in fruitfulness; a survey of the petiole nutrient status in 30 tropical vineyards showed a significant negative correlation ($r = -0.946$) between petiole N and fruitfulness (Muthukrishnan and Srinivasan, 1974).

Optimum phosphorus nutrition promotes bud fruitfulness by increasing vine vigor (Kobayashi, 1961); phosphate deficiency is detrimental to inflorescence formation. Low N, high P and water stress are the factors associated with high fertility in Sultana vines (Baldwin, 1966). Under tropical conditions, petiole P content is positively correlated with the yield of grapes (Muthukrishnan and Srinivasan, 1974). P is present in xylem sap in mineral form at the beginning of the so-called 'bleeding period' but it is mostly in an organic form during flower differentiation and bud burst (Kirikoi, 1973).

There have been several suggestions for a role for potassium in inflorescence formation in the grapevine. Soil application of K in K-deficient vineyards in Michigan and in the Niagara Peninsula caused a marked increase in the fruitfulness of latent buds of Concord (Larsen, 1963). Similar effects of K-nutrition were found in Thompson Seedless vines in California (Christensen, 1975).

The growing of vine seedlings in mineral nutrient solutions induces precocious flowering; hydroponic culture forms part of a breeding strategy for the grapevine proposed by Bouquet (1977). Optimum levels of

N, P and K are associated with maximum cytokinin production by grape roots (Jako, 1976).

THE REGULATION OF FLOWERING: CONCLUSIONS

The effects of exogenous gibberellin and cytokinin on flowering are unequivocal and indicate that flowering is regulated by a gibberellin–cytokinin interaction. There is little information on the origin and action of endogenous phytohormones in flowering in the intact grapevine, but the available evidence is consistent with the concept of a gibberellin–cytokinin interaction. The xylem sap of the grapevine contains gibberellin and has a high cytokinin activity during bud burst. There is a specific effect of roots on inflorescence development and it is likely that the root system is an important source of the growth substances which regulate flowering.

The external factors which promote flowering in the grapevine, such as short-term exposure to high temperature, high light intensity and optimum levels of soil moisture and macronutrients, are also factors that promote cytokinin biosynthesis in plants. Conversely, factors which depress flower formation, such as low light intensity, low temperature and water stress, have an inhibitory effect on endogenous cytokinin production. Moreover, exogenous cytokinins promote flower formation in grapevines grown at non-inductive low temperatures, and in plants grown in the dark (Srinivasan and Mullins, 1981a,b). It is clear that cytokinin is of central importance in the control of flowering in the grapevine.

Fruit growth and its regulation

FRUIT SET

Most grapevine cultivars are highly floriferous and inflorescences are formed at most nodes along the cane.[4] Depending on the cultivar, each latent bud contains up to three inflorescences and each inflorescence may contain a thousand flowers. Of these flowers, 70–80% normally fail to develop into mature fruits; they shrivel and drop off. The term 'fruit set' is used to describe the transformation of flowers into fruits. Fruit set may refer either to the percentage of flowers in an inflorescence which

---

[4] A notable exception is Thompson Seedless (Sultanina) in which the basal nodes on the cane are usually vegetative.

grow into fruits (commonly 20–30%) or to the physiological processes involved in the early stages of fruit growth.

In seeded cultivars, berry growth is initiated by pollination and fertilization. The special cases of parthenocarpy and stenospermocarpy were described earlier. Flowers that are unpollinated, or which fail to be fertilized, shrivel and die. In the remainder, there is resumption of cell division in the pericarp and the berries begin to grow. However, many of these developing berries are shed during the next two to three weeks by the process of abscission. Included are berries that had been properly pollinated and fertilized and which were in active growth. Fruit set in the grapevine is completed within two to three weeks of anthesis and the fruitlets retained will normally develop to maturity. There are no further well-defined periods of fruit abscission, as are found with pome fruits, but abscission of grape berries may occur as a result of unfavorable environmental conditions such as water stress.

To summarize, the constituent processes of fruit set are (i) resumption of cell division in the pericarp and (ii) prevention of abscission-layer activation at the base of the pedicel. These two processes seem to be independently controlled. Fruits drop off while in active growth, and immature berries may be retained by a bunch without completing their normal growth and development. The latter is characteristic of the disorder known as 'hens and chickens' or 'coulure'.

Stimulation of fruit set is a ready means of increasing yield, and this has prompted much research on the use of plant growth regulators in viticulture (Considine, 1983). With regard to endogenous regulation, current concepts are not greatly different from those of 20 years ago (Mullins, 1967; Possingham, 1970; Coombe, 1973). Fruit set is subject to influence by each of the main groups of naturally occurring growth substances; there are reports of promotion of fruit set by applied auxins, cytokinins, gibberellin and various growth retardants. There are complex interactions of genotype, environment and growth regulators, which make it difficult to ascribe specific roles to individual compounds. A unifying hypothesis is that growth substances, endogenous or exogenous, influence fruit set through effects on the partitioning of organic nutrients (Coombe, 1973). This is consistent with promotion of fruit set by pinching, tipping or removal of young leaves, that is, treatments which reduce the numbers of competing sinks (Coombe, 1962). Promotion of fruit set by metabolite diversion is also in accord with the phenomenon of hormone-directed transport in which it is hypothesized that hormones originating in the seeds or pericarp direct the import of nutrients from other parts of the plant.

The available evidence supports the view that fruit set in grapes is subject to correlative control, but there are still some unanswered questions. For example, it is possible in Cabernet Sauvignon grapes to achieve a substantial improvement in fruit set by providing ideal conditions for pollination and fertilization and by reducing the number of competing sinks, but it has not been possible so far to obtain a level of fruit set in excess of 65%. Even under the most favorable circumstances there is a considerable wastage of apparently functional flowers, and this suggests that there may be differences in the potential of flowers to be transformed into fruits. According to Luckwill (1957) fruit development commences at the time of formation of the gynoecial primordium and factors operating during the stage of flower differentiation may predetermine, to a greater or lesser extent, the capacity of grape flowers to grow into berries.

BERRY GROWTH

Regardless of the mechanism of fruit set, berry enlargement has a double sigmoid pattern (Fig. 5.2). There are three arbitrary stages: I, the

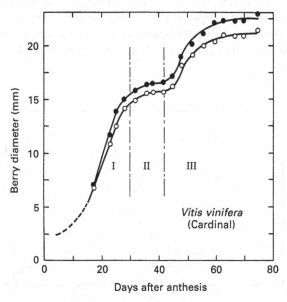

Fig. 5.2. The developmental pattern of fruit diameter (*V. vinifera* L. cv. Cardinal). Serial measurements of berry diameter were made on each of two berries (open and closed circles) for 70 days after anthesis. I, II and III refer to the three stages of berry growth (see text for details). From Matthews *et al.* (1987). Reproduced with permission

initial phase of rapid growth; II, the so-called lag phase of slow or no growth; and III, the final phase of resumed growth and maturation. The lag phase may be less distinct in seedless berries than in seeded berries (Iwahori *et al.*, 1968). Berries increase in mass about 4000-fold from anthesis to ripeness (Coombe, 1976). Growth is due to both cell division and cell enlargement, but osmotically driven cell enlargement of the pericarp cells is the main component. Cell number in the berry during the course of its development increases 3- to 4-fold, but cell volume increases approximately 300-fold. During development, pericarp volume increases from 10 to 20% of berry volume at anthesis to about 65% at maturity.

Once set has occurred, the berry increases rapidly in size and mass. Stage I is characterized by growth of the seed and pericarp, but there is little development of the embryo. Cell division in the pericarp ceases within three weeks of anthesis (Fig. 5.2) and is followed by the phase of cell expansion. The cessation of cell division proceeds from the placental tissue outwards to the epidermis. During Stage I, the green, hard, berries accumulate organic acids, and these are commonly measured as 'titratable acidity', TA (Fig. 5.3). The duration of Stage I is typically 40–60 days.

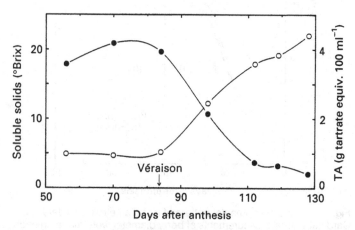

Fig. 5.3. Soluble solids (open circles) and titratable acidity (filled circles) of juice from berries of Cabernet franc. From Matthews and Anderson (1988). Reproduced with permission

STAGE II

Stage II is characterized by slow growth of the pericarp and by maturation of the seeds. Chlorophyll content and the rates of photosynthesis and respiration decrease. Titratable acidity reaches a maximum of approximately 0.27 M (Fig. 5.3). Although general metabolism slows, embryo development is rapid. The berry remains a hard green organ until the end of Stage II. The lag phase lasts 7–40 days. The length of the lag phase determines whether a cultivar is early- or late-maturing.

STAGE III

The onset of Stage III is marked by softening of the berry and by color change in pigmented cultivars. The stage at which anthocyanin pigments first appear in a grape berry is known as *véraison*. The resumption of rapid growth during this period is due solely to cell expansion. The berry attains its maximum size and ripens during Stage III. The extent of fruit expansion that occurs during Stages I and III is dependent upon the cultivar. The titratable acidity decreases and a massive accumulation of hexose sugars occurs, which attains a concentration of greater than 1 M (Fig. 5.3). Stage III lasts approximately 35–55 days.

CONTROL OF FRUIT EXPANSION

The physiological basis of the double sigmoid growth pattern of the grape berry is not well understood. Nitsch (1953) suggested that competition for assimilates between the endocarp and expanding mesocarp limits fruit expansion during the lag phase. Although this is possible in the fruit of *Prunus* sp., where lignification of the endocarp occurs during Stage II, it cannot explain the growth habit of grapes. Contrary to some reports (Winkler *et al.*, 1974; Peynaud and Ribéreau-Gayon, 1971; Lavee and Nir, 1986), there is no lignification (hardening) of the endocarp in grape berries. Indeed, an endocarp is generally not a distinguishable part of the fruit wall in the grape. Another hypothesis is that fruit growth is controlled by phytohormones produced in the seed (Chalmers and van den Ende, 1977).

Seeds are a rich source of growth substances, which diffuse into the surrounding tissues, but it is unlikely that seeds play a major role in the growth pattern of *Vitis* fruits because the phenological relationship between seed and pericarp development is highly variable among seeded cultivars (Peynaud and Ribéreau-Gayon, 1971). Seeds become hardened well before fruit maturity in late-maturing cultivars, but they remain immature at harvest in early-maturing cultivars. Also, berries of seedless cultivars, whether development was parthenocarpic or steno-spermocarpic, exhibit a double sigmoid growth pattern which is similar to that of seeded cultivars.

The biophysical factors known to regulate cell and organ expansion may provide a more satisfactory explanation of the double sigmoid growth pattern of the grape berry. It is likely that the resumption of growth in Stage III is attributable to an increase in turgor, the water potential gradient for water uptake (the driving force for cell expansion), and to an increase in the extensibility of the cell walls (Lockhart, 1965). The accumulation of sugars during Stage III leads to a major decrease in the solute (osmotic) potential of the berry (Fig. 5.4). This implies that the water potential difference between the source xylem and the expanding pericarp cells increases at veraison. Thus, the rate of berry expansion is likely to increase at this stage because there is an increase in the driving force for water uptake.

There is evidence of changes in the characteristics of cell walls of grape berries at véraison. Berry firmness decreases rapidly a few days

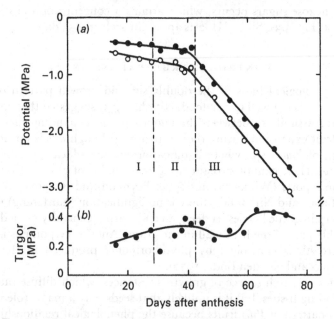

Fig. 5.4. Water relations of Cardinal berries during development. I, II and III refer to growth stages (see text). (a) Water potential (filled circles) and solute potential (open circles); (b) turgor. From Matthews *et al.* (1987) Reproduced with permission

before the resumption of growth (Coombe and Bishop, 1980). The spherical shape of the berry dictates that the stresses imposed on cell walls are greatest for the outer cell walls. Berry expansion appears to be limited by the capacity for expansion of cells of the dermal system. Berry cracking occurs in the field, and peeled berries expand at rates two times greater than intact controls. These observations indicate that the mesocarp is capable of expanding at significantly greater rates than the skin. Matthews et al. (1987) found that the resumption of growth was temporally correlated with a rapid increase in the plastic extensibility of isolated berry dermal tissue (Fig. 5.5). This correlation suggests that the resumption in growth in Stage III is due to an increased yielding of the dermal wall to existing turgor.

Dye perfusion and photomicrographic studies indicate that the xylem vessels within the berry are physically disrupted during the onset of Stage III. The appearance of discontinuities suggests that the vessels are torn apart during berry expansion (Findlay et al., 1987; Düring et al., 1978). Dermal cell walls become thin during Stage III (Considine and Knox, 1979) and the thinning of dermal walls, together with the disrup-

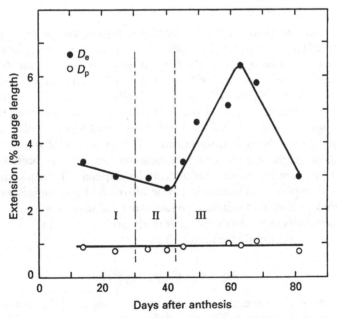

Fig. 5.5 Elastic (D_e) and plastic (D_p) extension of dermal strips of Cardinal berries. I, II and III refer to berry growth stages (see text). From Matthews et al. (1987). Reproduced with permission

tion of xylem, suggest that the biosynthesis of structural polymers may not resume with the resumption of berry growth. The second growth phase may represent an unusual type of growth in which wall synthesis declines but wall loosening continues. If fruit softening in grape berries proceeds in a manner similar to other fruit, it is likely that the cell walls may be altered by increased activity of wall-degrading enzymes such as polygalacturonase and cellulase.

Final berry size is dependent upon several factors under both genetic and environmental control. The number of pericarp cells, determined largely before anthesis and during Stage I, is an important determinant of potential volume. The number of cells is determined genetically and by environmental parameters, including temperature and water status. In seeded cultivars, seed number is positively correlated with fruit fresh mass, dry mass and accumulation of translocated photosynthate. The supply of photosynthate, determined by the amount of photosynthetic leaf area, the rate of photosynthesis, and by the presence of competing sinks, is a factor that can be exploited by cultural practices so as to maximize berry size.

GAS EXCHANGE

The developing berries are photosynthetic organs and produce phosphorylated sugar products (Fig. 5.6). The significance of fruit photosynthesis to the carbon budget of the berry is not clear, but its contribution is probably small. The rate of fruit photosynthesis seldom exceeds the CO_2 compensation point (Frieden et al., 1987). The rate of respiration reaches a maximum early in berry development and decreases thereafter (Figs. 5.6 and 5.7). Grape berries lose small amounts of water to the atmosphere through transpiration. The frequency of stomata and lenticels on grape berries is low, and these openings become occluded by deposits of epicuticular wax during fruit development. Transpiration by berries in response to changes in temperature and light diminishes during development, and diffusive conductance for water transport to the atmosphere decreases from approximately 20×10^{-3} cm s^{-1} for green fruit to 6×10^{-3} cm s^{-1} for mature fruit.

FRUIT RIPENING

The absence of a large, transient, increase in respiratory CO_2 evolution at the onset of ripening (Fig. 5.7) defines the grape as a *non-climacteric* fruit. The commencement of ripening in non-climacteric fruit is often less marked than that of climacteric fruit but the onset of ripening in the

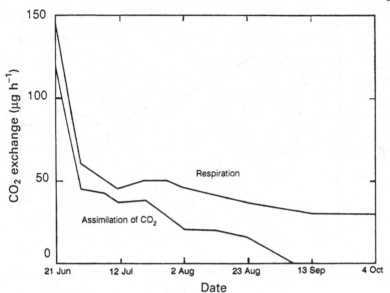

Fig. 5.6. Changes in photosynthetic CO_2 assimilation and respiratory rate (CO_2 output) during development and maturation of grapes. Rates are on a per berry basis. From Geisler and Radler (1963). Reproduced with permission

grape berry is very distinct. The transition from Stage II to Stage III marks the beginning of berry ripening and includes many physiological changes (listed below). Most of these changes occur rapidly, i.e. within 24–48 h:

Softening of the berry
Increase in rates of expansion and dry mass gain
Accumulation of hexoses
Decrease in titratable acidity
Decrease in malate
Increase in pH of cell sap or grape juice
Increase in respiratory quotient (RQ)
Onset of anthocyanin synthesis in colored cultivars (véraison)
Increase in concentrations of proline and arginine

The term véraison is applied to the onset of fruit ripening in grape. Originally referring only to the color change, its use has become more general in recent years, and it now refers to both softening and the resumption of growth. It should be noted that grape berries do not continue to ripen after removal from the rachis.

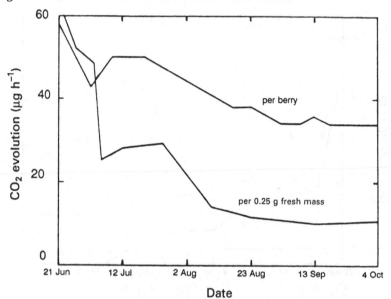

Fig. 5.7. Changes in the respiration rate of the grape during development. The rate of respiration is expressed on a per berry and fresh mass basis. From Geisler and Radler (1963). Reproduced with permission

CHANGES IN PHYTOHORMONES ASSOCIATED WITH RIPENING

Fruit ripening is considered to be a senescence process under hormonal control, but the roles of phytohormones in grape berry ripening are not clear. Müller-Thurgau (1898) showed that berry size was dependent upon seed number. Although differences of seed number are likely to produce differences in hormone production, such differences among grape berries with different numbers of seeds have not been established (Cawthon and Morris, 1982; Scienza *et al.*, 1978).

As a non-climacteric fruit, there is no burst of endogenous ethylene production by the ripening grape berry, nor is there a consistent growth response to exogenous ethylene. Ethylene evolution remains low throughout grape maturation; treatments that increase ethylene evolution do not accelerate ripening (Coombe and Hale, 1973; Inaba *et al.*, 1976). Ethylene may be involved in color development because application of ethephon (2-chloroethylphosphonic acid), a source of ethylene, causes increased anthocyanin accumulation in colored cultivars.

Gibberellin (Coombe, 1960; Ito *et al.*, 1969; Inaba *et al.*, 1976) and cytokinin (Inaba *et al.*, 1976) activities were high early in Stage I, but decreased to low levels before the onset of ripening. In seedless cultivars, the sensitivity of berry growth to exogenous GAs suggests a role for this class of phytohormone in the endogenous regulation of berry development. Although seeded cultivars respond little to exogenous GA, higher concentrations of endogenous GA and larger berries are usually found in seeded than in seedless cultivars (Iwahori *et al.*, 1968).

Decreases in auxin concentration and increases in ABA concentration are correlated with the initiation of berry ripening (Coombe and Hale, 1973; Cawthon and Morris, 1982; Inaba *et al.*, 1976; Scienza *et al.*, 1978). In general, berry growth is insensitive to exogenous auxin; an exception is 4-chlorophenoxyacetic acid, which enhances growth of seedless berries. Applications of ABA can accelerate sugar accumulation, and applications of growth regulators that alter ripening also increase the endogenous ABA concentrations.

These observations, and the close association between increased ABA concentrations and the resumption of growth, have led to the suggestion that ABA triggers ripening in grapes (Coombe and Hale, 1973). It is possible that high concentrations of ABA may be produced by the seed to inhibit precocious germination of the maturing embryo. In one study the concentration of ABA in the seed was high early during Stage I, but decreased and remained low during Stages II and III. Also, there is evidence of ABA translocation to berries from leaves (Düring *et al.*, 1978). The concentration of ABA in the skin is closely correlated with the concentration of sugars.

CARBOHYDRATES AND ORGANIC ACIDS

The concentration of sugars increases while that of organic acids decreases during ripening (see Fig. 5.3). As acidity decreases, the pH of pericarp sap increases. The concentration of sugars at maturity greatly exceeds that of organic acids in Stage II, indicating that the change in concentration of sugar is not attributable to the conversion of accumulated organic acids to sugars. The accumulation of sugars is dependent upon leaf photosynthesis and on the import of sucrose, but it is not due to a rapid increase in photosynthetic activity. The accumulation of sugars by the berry represents a significant shift in translocation patterns. Before véraison, shoot apices are the major sinks. Berries accumulate high concentrations of hexoses, which are translocated from proximal leaves on the bearing shoot and, to a lesser extent, from storage tissues. Invertase activity increases at véraison, induced perhaps by the increased influx of substrate sucrose.

The organic acids in berries are dominated by tartrate and malate:

and these two acids account for over 90% of the total titratable acidity. In comparison with other fleshy fruits, grape berries accumulate high concentrations of tartaric acid and low concentrations of citric acid. Biosynthesis of tartaric acid and malic acid occur via different pathways. The decrease in acid concentration during the ripening of grape berries may be due to at least four different mechanisms.

(1) Dilution by increase in berry volume. Catabolism of tartrate occurs at a relatively slow rate throughout berry development. Expressing amounts of malic and tartaric acids on a 'per berry' basis demonstrates that there is no decline in tartrate but a rapid reduction in the total amount of malate. Thus, decreases in the concentration of tartaric acid during ripening are apparently due to dilution, but malate content per berry decreases significantly, and the decrease in berry acidity is due primarily to the decrease in malate content.

(2) Activation of acid breakdown. The respiratory quotient changes from 1.0 before véraison (indicating carbohydrates as the energy source), to about 1.4 (indicating carboxylic acids as the energy source) during ripening.

(3) Inhibition of synthesis. The malate in green berries is probably synthesized from translocated sucrose, a conclusion that is supported by lack of photosynthetic self-sufficiency of the berry throughout most of its development. If malate synthesis occurs via sucrose and is inhibited after véraison, it follows that there is altered metabolism of exogenous sucrose. The capacity to synthesize malate from exogenous sucrose decreases greatly at véraison (Possner et al., 1983).

(4) Transformation from acid to sugar. After véraison, there is gluconeogenic carbon flux in the berry. Evidence for this includes the rapid evolution of CO_2 from labeled sucrose in immature berries, but very low evolution from mature fruit (Takimoto et al., 1976); the rapid conversion of labeled malate to glucose in ripening fruit as distinct from the rapid conversion of labeled glucose to malate in immature fruit; and the rapid decline in concentration of oxaloacetate, malate, and citrate after véraison. These observations indicate that the primary direction of carbon flux in glycolysis is different before and after véraison. At the time of transition, there is an increase in the activities of invertase, sucrose synthase, sucrose phosphate synthase, and sucrose phosphatase (Hawker, 1969).

Ruffner et al. (1984) have argued that malic enzyme (ME) plays a regulatory role in malic acid degradation. Maximum ME activity occurs when malate concentration is declining rapidly (Fig. 5.8). Although the ME-catalyzed reaction is partly reversible in vitro, and some ME activity is present even during the period when malate is accumu-

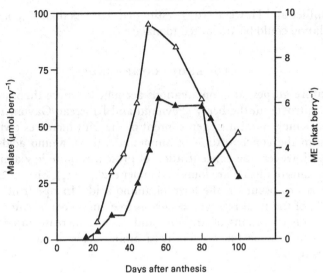

Fig. 5.8. Developmental changes in malic acid content (open triangles) and malic enzyme activity (closed triangles) in Riesling and Sylvaner berries. From Ruffner *et al.* (1984). Reproduced with permission

lating, ME does not convert exogenous labeled precursor to malate in the berry. Furthermore, respiratory degradation of labeled malate increases from 37% in the green berry to about 90% in ripening fruit (Steffan and Rapp, 1979). The activity of malic enzyme is highest in the center of the grape berry during seed development, but highest ME activity is found in the peripheral tissues during fruit ripening (Possner *et al.*, 1983).

The transient increase in ABA that occurs at the onset of ripening may play a role in the transition of gluconeogenesis. Treatment of excised berry slices from unripe fruit with ABA increased gluconeogenic activity and the activity of several gluconeogenic enzymes including malate dehydrogenase (Palejwala *et al.*, 1985). When cycloheximide was added to the incubation, the activities of gluconeogenic enzymes were unaltered, suggesting that protein synthesis was required for the ABA to be effective.

The extent to which gluconeogenesis contributes to the increase in sugar concentration of grape berries is difficult to estimate. It is clear that the increase in sugar during ripening cannot be due solely to the gluconeogenic flux of organic acids because the pool of acid is inadequate to produce the high concentrations of sugars observed in ripe fruit. Ruffner and Hawker (1977) estimated that less than 5% of sugar accumulation could be attributed to malate.

NITROGENOUS COMPOUNDS

In immature grapes, ammonia cations account for more than 50% of the total nitrogen in the berry (Peynaud and Ribéreau-Gayon, 1971). During ripening, total nitrogen content of the fruit increases owing to increases in the concentration of ammonia cations, amino acids and proteins. However, the concentration of protein is quite low and only trace amounts of nitrate are found (Winkler *et al.*, 1974). Most nitrogen in the pericarp occurs in the form of amino acids. In ripe fruit, more than 50% of the total nitrogen occurs as free amino acids. After véraison, the concentrations of arginine and proline increase rapidly (2- to 6-fold) to become the predominant amino acids in most cultivars (Kliewer, 1970).

PHENOLICS

Phenolic compounds in grapes comprise derivatives of hydroxycinnamic acid, including caffeic acid and coumaric acid, flavonoids, including anthocyanins (which are discussed below), flavonols, and tannins (complex esters of phenolics, acids and sugars). Non-flavonoid phenolics accumulate primarily in the vacuoles of mesocarp cells, but flavonoids accumulate in the dermal cells.

Caftaric and coumaric acids are the major phenolic substances in unmodified grape juice (Singleton *et al.*, 1986). The capacity to synthesize each of these compounds is maintained throughout ripening. Phenolics are subject to rapid oxidation; *S*-glutathionylcaftaric acid has been shown to function as an indicator of the degree of enzymic oxidation in berries (Singleton *et al.*, 1985).

The shikimic acid pathway functions in berries and is the likely source of cinnamates and anthocyanins, although other pathways may contribute other phenols (Peynaud and Ribéreau-Gayon, 1971). Control points and the signal for the onset of anthocyanin biosynthesis in grapes are unknown, but the activity of phenylalanineammonia lyase (PAL) is a common control point in many plants. In grape berries, the concentration of total phenols in the skin of colored cultivars decreases until véraison, but it then increases slightly during anthocyanin accumulation. In one study, PAL activity is shown to decrease from very high levels during Stage I in white, purple, and black cultivars; however, near the onset of véraison PAL activity increased rapidly for the colored cultivars and this was preceded by an increase in skin sugar concentration (Kataoka *et al.*, 1983). The concentration of free ABA also increases near the onset of véraison.

PIGMENTS

Green berries contain significant amounts of chlorophyll early in their development, but the concentration declines steadily until the green color is obscured by other pigments. Also present are carotene and xanthophylls, which give the white or yellow cultivars their color. In red or black cultivars, anthocyanins are responsible for the color. Pigments are generally confined to the vacuoles of a few cell layers immediately below the epidermis, but there are a few cultivars that contain pigment in the mesocarp cells. These cultivars are described as 'teinturiers', the best known of which is Alicante Bouschet.

The parent compounds of the phenolic pigments of the grape berry are the anthocyanidins (cyanidin, delphinidin, petunidin, peonidin and malvidin). These compounds are modified by attachment of glucose moieties to form anthocyanins. The principal anthocyanin pigments of red and black *vinifera* grapes are monoglucoside anthocyanins, predominantly malvidin 3-monoglucoside. *Vitis* species other than *vinifera* contain both monoglucosides and diglucosides, and the presence of diglucosides has been used to detect frauds in which high-priced *vinifera* wines have been adulterated with wines from high-yielding American hybrids or 'producteurs directs'.

Anthocyanins are red when acidic, colorless at pH near 4.0, and purple at pH above about 4.5. Under alkaline conditions, a blue color can be produced. The common anthocyanins in *Vitis* species are shown in Table 5.1. The causes of the blue color of some cultivars is not known. The blue color of other plant tissues has been attributed to anthocyanin complexes with alkaline metals or to co-pigmentation in anthocyanin–flavonoid complexes.

The accumulation of anthocyanins and of sugar in the ripening berry are closely associated. Although earlier work failed to demonstrate a consistent relation between anthocyanin concentration and the concentration of sugars in the entire fruit, the progression of anthocyanin synthesis in berry dermal systems is highly correlated with hexose concentration (Pirie and Mullins, 1977). Work with isolated leaf and fruit tissue suggests that endogenous sugars act as a trigger for the accumulation of anthocyanin and other phenolic compounds (Pirie and Mullins, 1976). The synthesis of anthocyanins may continue steadily from after véraison until maturity or, under some conditions, the accumulation

Table 5.1. *Phenolic compounds of grapes*

General formula	Nature of specific compounds	Type of combination
Benzoic acids	$R=R'=H$ gives *p*-hydroxybenzoic acid $R=OH, R'=H$ gives protocatechuic acid $R=OCH_3, R'=H$ gives vanillic acid $R=R'=OCH_3$ gives syringic acid $R=H$ gives salicylic acid $R=OH$ gives gentisic acid	Combinations labile to alkali (esters and other compounds)
Cinnamic acids	$R=H$ gives *p*-coumaric acid $R=OH$ gives caffeic acid $R=OCH_3$ gives ferulic acid	Acyl combinations on anthocyanins' sugar on tartaric acid and caftaric, coutaric, fertaric acids
Flavonols	$R=R'=H$ gives kaempferol $R=OH, R'=H$ gives quercetin $R=R'=OH$ gives myricetin	Two or three glycosides are present, and one glucuronoside
Anthocyanidins	$R=OH, R'=H$ gives cyanidin $R=OCH_3, R'=H$ gives peonidin $R=R'=OH$ gives delphinidin $R=OCH_3, R'=OH$ gives petunidin $R=R'=OCH_3$ gives malvidin	3-Glucosides and acylated glucosides and additional forms depending on the species of *Vitis*
Tannin 'precursors'	$R=OH, R'=H$ gives catechin $R=R'=OH$ gives gallocatechin $R=OH, R'=H$ gives leucocyanidin $R—R'=OH$ gives leucodelphinidin	Tannins present are polymers of flavans, chiefly flavan-3,4-diols. These flavans are present in small amounts as monomers

of anthocyanin may occur chiefly during the first half of the ripening period.

Aroma compounds are synthesized late in berry development and are largely restricted to the dermal tissue. Accumulation of aroma compounds is not closely correlated with the sugar concentration. Several hundred volatile compounds have been identified in ripe grapes; they typically occur at concentrations of much less than one milligram per kilogram of fruit. Few of these compounds have been studied extensively because they are present at extremely low concentration, and because they have a tendency to oxidize or break down during isolation. Elaborate analytical techniques are required for this research.

Methyl anthranilate is the volatile substance responsible for the distinctive 'foxy' aroma of *Vitis labrusca*. In Muscats, terpene alcohols, including linalool, play a major role in determining the fruit and wine aroma. Linalool is first detected more than two weeks after the inception of ripening. Other volatiles, including terpineol and geraniol, were detected two to four weeks later (Hardy, 1970). In Riesling, linalool and linolenic acid are present and appear to determine the characteristic aroma of the must and the wine. Cabernet Sauvignon contains low concentrations of 2-methoxy-3-isobutylpyrazine. This compound has a strong capsicum or green bell pepper aroma, and it is associated with the so-called vegetative character of some wines made from Cabernet Sauvignon (Bayonove *et al.*, 1975).

Fruit quality

TABLE GRAPES AND RAISINS

Visual factors are very important in determining the quality of table grapes. These factors include berry size, color, and the conditions of the wax bloom on the fruit surface. The fruit is harvested at a lower concentration of sugars for table grapes than for wine. Also, the berries of table grape cultivars have a more firm or crispy texture than wine grapes. For raisins, seedlessness, size and a high sugar content are major quality factors.

WINE GRAPES

Quality might be defined as those attributes of the grape that make it attractive or pleasant to eat as a fresh fruit or to drink as wine (Webb,

1981). The absolute concentrations of sugars and acids, as well as their ratio, play important roles in the flavor of grapes. After fermentation, the sugar is converted to alcohol; this determines whether the wine tastes thin and watery or alcoholic and hot. Excessive acidity produces wines that are too tart, but grapes that are deficient in acid produce wines that have a flat and uninteresting taste. Acidity is also involved in the inhibition of oxidation and spoilage, and it is an important factor in wine stability. In this regard, there is a positive correlation between juice pH and the concentration of potassium (Boulton, 1980).

Phenolics determine color in nearly all wines and they are major factors in the flavor of red wines. Phenolics contribute a desirable bitterness and astringency to wine flavor (Webb, 1981). However, excessive tannin, extracted from seeds, pulp, and skin, results in a wine that is astringent and which has an unpleasant mouth-feel.

The amino acids serve as important nitrogen sources for the developing yeast during fermentation, but proline is not degraded under the usual anaerobic conditions (Cooper, 1982). Excessive nitrogen fertilization may lead to high concentrations of arginine in berries with the possible consequence of producing urethane, a carcinogen, in the resultant wine (Ough et al., 1989, 1990).

Aroma compounds that are important in wine flavor arise from both the fruit and as products of fermentation. In general, the berry-derived aroma compounds are more important for those cultivars in which distinct aromas are associated with the fruit. For white cultivars, most aroma compounds originate during fermentation. No single compound seems to be responsible for the characteristic flavor of any grape cultivar.

Environmental control of fruit growth and composition

EFFECTS OF TEMPERATURE

In cool regions, where ripe fruits may not contain adequate sugar concentrations for winemaking, quality is often defined by sugar level (e.g. German Mosel). In hot regions, ripening occurs more quickly but low acidity in the juice is a frequent problem, resulting in wines that are bland and of low quality. In addition, significant oxidation reactions, especially those involving phenolics, occur at elevated temperatures.

The rate of fruit development is not a simple function of temperature. The extensive work of Hale and Buttrose (1974) with Cabernet Sauvignon grown in growth chambers showed that the sensitivity of berry development to temperature changed with stage of development.

Vines were grown under three temperature regimes (18/13, 25/20, 35/30) and permutations thereof. When grown under continuous moderate conditions Stage I was 46 days; Stage II, 15 days; and Stage III, 28 days. More than 66% of berry growth occurred during the first growth phase, Stage I. In Stage I, berry growth was an inverse function of temperature, but the high temperature regime greatly reduced growth. This effect was maintained through maturity. The duration of Stage I was inversely related to temperature. The length of the lag phase, Stage II, was least under moderate temperatures and lengthened dramatically by high temperatures. The duration of Stage II exhibited the greatest sensitivity to temperature, and it was variation in the length of Stage II which primarily determined the days required to mature the fruit. The duration of the ripening phase, Stage III, did not exhibit a clear dependence upon temperature although the duration of Stage III varied considerably among treatments. Somewhat surprisingly, the longest period required for ripening was under conditions of continually high temperature.

Lower growth temperatures had no effect on final fruit size, although véraison was delayed (Hale and Buttrose, 1974). At moderate temperatures there was a sharp transition in growth rate at the onset of Stage III as compared with the higher and lower temperature regimes. There was also a much more rapid onset of fruit coloring at the moderate temperatures. Color development was delayed most by the high temperature regime. High temperatures during Stage II and III generally decreased the Brix (a measure of soluble solids), but Stage I temperature appeared to have little effect. High temperatures in Stage III decreased Brix only when temperatures were also high in Stages I and II. Changes in temperature after véraison had only slight effects on fruit size and sugar accumulation, but there was a weak inverse relation between temperature during ripening and the final Brix. Acidity was low in treatments in which one or more stages of fruit growth was at high temperatures. Acidity was high when Stage I and II were at low or intermediate levels and when Stage III was at a low temperature. Therefore, the temperature regimes at times other than during fruit ripening also have major effects on fruit composition.

In general, Brix increases slightly with increasing temperature up to about 30 °C. Malate concentration is strongly dependent on temperature throughout the range 15–40 °C, being lower at the higher temperatures. The work of Kliewer (1970, 1973) showed that juice pH is positively correlated with ambient temperature, and that TA and malate concentrations of juice are negatively correlated with temperature. Night and day temperatures have similar effects but day temperature is the more important for most aspects of fruit ripening. The accumulation of proline is greatly affected by night temperatures (Kliewer, 1973).

EFFECTS OF LIGHT

Light is an important factor in anthocyanin synthesis, especially in certain cultivars, but other effects of light are much less clear than those of temperature. Low light intensity decreases berry size, Brix, pH, and the concentration of proline, and increases TA and the concentration of malate and arginine (Kliewer and Lider, 1970). When the incident radiation directly on the fruit has been manipulated, only slight differences in composition have been observed. For example, the harvested juice of exposed and shaded clusters of Cabernet Sauvignon grapes grown in Napa Valley exhibited no significant differences in Brix, tartrate, malate, potassium, pH, or phenolics (Crippen and Morrison, 1986). Removal of basal leaves in Sauvignon blanc (Bledsoe et al., 1988) resulted in very small increases in soluble solids and TA, and decreases in pH and malate. In these experiments, berry temperatures were higher in the high-intensity light treatments than in the shaded clusters. The extent to which the observed differences in fruit composition were due to light or temperature is not clear.

Work by Dokoozlian (1990) with Cabernet Sauvignon and Pinot noir in controlled environments showed that visible radiation enhances fruit growth, sugar accumulation and anthocyanin accumulation. These responses are saturated at very low intensities, approximately 1–10% of full sunlight. Such low light intensities occur under very dense vineyard canopies. Effects of light quantity on berry growth were observed primarily when shading occurred early in berry development. Shading during Stage III had no effect on growth or sugar accumulation but it decreased malate degradation and anthocyanin synthesis. Shading (early in the season) also results in a delay of véraison in some cultivars, including Cabernet Sauvignon and Pinot noir.

Recently, shading experiments with Cabernet Sauvignon indicated an effect of light intensity on berry growth, sugar accumulation, malate, TA, anthocyanins, total phenols and ammonia (Smart et al., 1985). The extent to which these responses were due to differences in the red : far-red ratio is not clear from these experiments because the photon flux density (total incident radiation) was varied at the same time as the incident red: far-red radiation. It is noteworthy that some important enzymes in fruit ripening are phytochrome-regulated, e.g. phenylalanine ammonia lyase; malic enzyme and these enzymes may be light-regulated in grape berries.

Literature cited

Alleweldt, G. 1963. Einfluss von Klimafactoren auf die Zahl der Infloreszenzen bei Reben. *Wein. Wiss.* **18**: 61–70.

Alleweldt, G. 1964. Die Beeinflussing der Ertragsbildung bei Reben durch Tageslänge und Temperatur. *Kali-Briefe* (Fachgebiet 5), series 1, pp. 1–12.

Alleweldt, G. and Hofacker, W. 1975. Influence of environmental factors on bud burst, flowering, fertility and shoot growth of vines. *Vitis* **14**: 103–15.

Baldwin, J.G. 1964. The relation between weather and fruitfulness of the Sultana vine. *Aust. J. Agric. Res.* **15**: 920–8.

Baldwin, J.G. 1966. The effect of some cultural practices on nitrogen and fruitfulness in the Sultana vine. *Am. J. Enol. Vitic.* **17**: 58–62.

Bayonove, C., Cordonnier, R. and Dubois, P. 1975. Etude d'une fraction caractéristique de l'arome du raisin de la variété Cabernet Sauvignon; mise en évidence de la 2-methoxy-3-isobutylpyrazine. *C. R. Acad. Sci. Paris D* **281**: 75–8.

Bledsoe, A.M., Kliewer, W. M. and Marois, J. J. 1988. Effects of timing and severity of leaf removal on yield and fruit composition of Sauvignon blanc grapevines. *Am. J. Enol. Vitic.* **39**: 49–54.

Boulton, R.B. 1980. The general relationship between potassium, sodium and pH in grape juice and wine. *Am. J. Enol. Vitic.* **31**: 182–6.

Bouquet, A. 1977. Amélioration génétique de la vigne: Essai de définition d'un schéma de sélection applicable à la création de nouvelles variétés. *Ann. Amélior. Plantes* **27**: 75–86.

Braikov, D. 1975. The effect of temperature on the duration of organogenic phases in grapevines. *Gradinar. Lozar. Nauka* **12**: 108–14.

Buttrose, M.S. 1968. Some effects of light intensity and temperature on dry weight and shoot growth of grapevines. *Ann. Bot.* **32**: 753–65.

Buttrose, M.S. 1969a. Fruitfulness in grapevines: Effects of light intensity and temperature. *Bot. Gaz.* **130**: 166–73.

Buttrose, M.S. 1969b. Fruitfulness in grapevines: Effects of changes in temperature and light regimes. *Bot. Gaz.* **130**: 173–9.

Buttrose, M.S. 1969c. Vegetative growth of grapevine varieties under controlled temperature and light intensity. *Vitis* **8**: 280–5.

Buttrose, M.S. 1970a. Fruitfulness in grapevines: Development of leaf primordia in buds in relation to bud fruitfulness. *Bot. Gaz.* **131**: 78–83.

Buttrose, M.S. 1970b. Fruitfulness in grapevines: the response of different cultivars to light, temperature and day length. *Vitis* **9**: 121–5.

Buttrose, M.S. 1974a. Climatic factors and fruitfulness in grapevines. *Hort. Abstr.* **44**: 319–25.

Buttrose, M.S. 1974b. Fruitfulness in grapevines: effect of water stress. *Vitis* **12**: 299–305.

Cawthon, D.L. and Morris, J.R. 1982. Relationship of seed number and maturity to berry development, fruit maturation, hormonal changes, and uneven ripening of 'Concord' (*Vitis labrusca* L.) grapes. *J. Amer. Soc. Hort. Sci.* **107**: 1097–104.

Carr, D.J. 1968. The relationship between florigen and the flower hormones. *Ann. N.Y. Acad. Sci.* **144**: 305–12.

Chailakhyan, M. 1977. Hormonal regulation of plant flowering. *Plant growth regulation.* (ed. P.E. Pilet) (*Proc. 9th Int. Conf. Plant Growth Substances*), pp. 258–72. Springer-Verlag, Berlin.

Chalmers, D.J. and van den Ende, B. 1977. The relation between seed and fruit development in the peach (*Prunus persica* L.) *Ann. Bot.* **41**: 707–14.

Christensen, P. 1975. Long-term responses of Thompson Seedless vines to K-fertilizer treatment. *Am. J. Enol. Vitic.* **26**: 179–83.

Considine, J. A. 1983. Concepts and practice of use of plant growth regulating chemicals in viticulture. In *Plant growth regulating chemicals* (ed. L.G. Nickell), vol. 1, pp. 89–193. CRC Press, Boca Raton, Florida.

Considine, J.A. and Knox, R.B. 1979. Development and histochemistry of the cell walls and cuticle of the dermal system of the fruit of the grape, *Vitis vinifera* L. *Protoplasma* **99**: 347–65.

Coombe, B.G. 1962. The effects of removing leaves, flowers and shoot tips on fruit set in *Vitis vinifera* L. *J. Hortic. Sci.* **37**: 1–15.

Coombe, B.G. 1967. Effect of growth retardants on *Vitis vinifera*. *Vitis* **6**: 278–87.

Coombe, B.G. 1970. Fruit set in grapevines: The mechanism of the CCC-effect. *J. Hort. Sci.* **45**: 415–25.

Coombe, B.G. 1973. The regulation of set and development of the grape berry. *Acta Hortic.* **34**: 261–73.

Coombe, B.G. 1976. The development of fleshy fruits. *Ann. Rev. Plant Physiol.* **27**: 507–28.

Coombe, B.G. and Bishop, G.R. 1980. Development of the grape berry. II. Changes in diameter and deformability during veraison. *Aust. J. Agric. Res.* **31**: 125–35.

Coombe, B.G. and Hale, C.R. 1973. The hormone content of ripening grape berries and the effects of growth substance treatments. *Plant Physiol.* **51**: 629–34.

Cooper, T.G. 1982. Nitrogen metabolism in *Saccharomyces cerevisiae*. In *The molecular biology of the yeast Saccharomyces: metabolism and gene expression* (ed. J.N. Strathern, E.W. Jones and J.R. Broach), pp. 33–99. Cold Spring Harbor Laboratory, Cold Spring Harbor, New York.

Crippen, D.D. and Morrison, J.C. 1986. The effects of sun exposure on the compositional development of Cabernet Sauvignon berries. *Am. J. Enol. Vitic.* **37**: 235–242.

Dokoozlian, N.K. 1990. *Light quantity and light quality within Vitis vinifera L. grapevine canopies and their relative influence on berry growth and composition.* Ph.D. thesis, University of California, Davis. 324 pp.

Düring, H. and Alleweldt, G. 1973. Der Jahrengang Abscisinsäure in Vegetativen Organen von Reben. *Vitis* **12**: 26–32.

Düring, H., Alleweldt, G. and Koch, R. 1978. Studies on hormonal control of ripening in grape berries. *Acta Hort.* **80**: 397–405.

Evans, L.T. 1971. Flower induction and the florigen concept. *Ann. Rev. Plant Physiol.* **22**: 365–94.

Findlay, N., Oliver, K.J., Nii, N. and Coombe, B.G. 1987. Solute accumulation by grape pericarp cells. IV. Perfusion of pericarp apoplast via the pedicel and evidence for xylem malfunction in ripening berries. *J. Exp. Bot.* **38**: 668–79.

Frieden, K.-H., Lenz, F. and Becker, H. 1987. Der CO_2-Gaswechsel von Traubenbeeren verschiedener Rebsorten. *Wein Wissensch.* **42**: 219–34.

Geisler, G. and Radler, F. 1963. Development and maturation of fruits in *Vitis*. *Berichte der Deutschen Botanischen Gesellschaft* **76**: 112–19.

Hale, C.R. and Buttrose, M.S. 1974. Effect of temperature on ontogeny of berries of *Vitis vinifera* L. cv. Cabernet Sauvignon. *J. Amer. Soc. Hort. Sci.* **99**: 390–4.

Hardy, P.J. 1970. Changes in volatiles of muscat grapes during ripening. *Phytochemistry* **9**: 709–15.

Hawker, J.S. 1969. Changes in the activities of malic enzyme, malate dehydrogenase, phosphopyruvate carboxylase and pyruvate decarboxylase during the development of a non-climacteric fruit (the grape). *Phytochemistry* **8**: 19–23.

Hopping, M.E. 1975. Effect of light intensity during cane development on subsequent bud break and yield of Palomino grapevines. *N.Z.J. Exp. Agric.* **5**: 287–90.

Huglin, P. 1960. Causes determinant les alterations de la floraison de la vigne. *Ann. Amélior. Plantes* **10**: 351–8.

Inaba, A., Ishida, M. and Sobajima, Y. 1976. Changes in endogenous hormone concentrations during berry development in relation to the ripening of Delaware grapes. *J. Jap. Soc. Hort. Sci.* **45**: 245–52.

Ito, H., Motomura, Y., Konno, Y. and Hatayama, T. 1969. Exogenous gibberellin as responsible for the seedless berry development of grapes. I. Physiological studies on the development of seedless Delaware grapes. *Tohoku J. Agric. Res.* **20**: 1–18.

Iwahori, S., Weaver, R.J. and Pool, R.M. 1968. Gibberellin-like activity in berries of seeded and seedless Tokay grapes. *Plant Physiol.* **43**: 333–7.

Jako, N. 1976. The relationship between nitrogen, phosphorous, potassium and magnesium nutrition and growth of grapevines and cytokinin production by the roots. *Szoleszet et Borazet* **1**: 35–47.

Kataoka, I., Kubo, Y., Sugiura, A. and Tomana, T. 1983. Changes in L-phenylalanine ammonia-lyase activity and anthocyanin synthesis during berry ripening of three grape cultivars. *J. Jap. Soc. Hort. Sci.* **52**: 273–9.

Kirikoi, Y.T. 1973. The effect of mineral fertilizers on changes in the composition of grapevine sap. *Hort. Abst.* **43**: 6748.

Kliewer, W.M. 1970. Free amino acids and other nitrogenous fractions in wine grapes. *J. Food Sci.* **35**: 17–21.

Kliewer, W.M. 1973. Berry composition of *Vitis vinifera* cultivars as influenced by photo- and nycto-temperatures during maturation. *J. Amer. Soc. Hort. Sci.* **98**: 153–9.

Kliewer, W.M. and Lider, L.A. 1970. Effects of day temperature and light intensity on growth and composition of *Vitis vinifera* L. fruits. *J. Amer. Soc. Hort. Sci.* **95**: 766–9.

Kobayashi, A.S. 1961. Effect of potassium and phosphoric acid on the growth, yield and fruit quality of grapes. 1. On the effect of the application level on non-fruiting and fruiting vines in sand culture. 2. On the effect of time and level of application in sand culture. *Mem. Res. Inst. Food Sci. Kyoto Univ.* **23**: 28–46.

Kobayashi, A., Yukinaga, H. and Nü, N. 1965. Studies on the thermal conditions of grapes. IV. Effect of day and night temperatures on the growth of Delaware. *J. Jap. Soc. Hort. Sci.* **34**: 77–84.

Kobayashi, A., Sigiura, A., Watanabe, H. and Yamamura, H. 1966. On the effects of daylength on the growth and flower bud formation of grapes. *Mem. Res. Inst. Food Sci. Kyoto Univ.* **27**, 15–27.

Larsen, R.P. 1963. Effect of potassium and magnesium fertilizers on the nutritional status and yield of a Concord grape vineyard. *Q. Bull. Mich. Agric. Exp. Sta.* **45**: 376–86.

Lavee, S. and Nir, G. 1986. Grape. In *Handbook of fruit set and development* (ed. S.P. Monselise), pp. 167–91. CRC Press, Boca Raton, Florida.

Lilov, D. and Andanova, T. 1976. Cytokinins, growth, flower and fruit formation in *Vitis vinifera*. *Vitis* **15**: 160–70.

Livne, A. and Vaadia, Y. 1972. Water deficits and hormone relations. In *Water deficits and plant growth* (ed. T.T. Kozlowski, vol. 3, pp. 241–75. Academic Press, New York.

Lockhart, J.A. 1965. An analysis of irreversible plant cell elongation. *J. Theor. Biol.* **8**: 264–75.

Loveys, B.R. and Kriedemann, P.E. 1973. Rapid changes in abscisic acid-like inhibitors following alterations in vine leaf water potential. *Physiol. Plant.* **28**: 476–9.

Luckwill, L.C. 1957. Hormonal aspects of fruit development in higher plants. *Symp. Soc. Exp. Biol.* **11**: 63–85.

Matthews, M.A., Cheng, G. and Weinbaum, S. A. 1987. Changes in water potential and dermal extensibility during grape berry development. *J. Amer. Soc. Hort. Sci.* **112**: 314–19.

Matthews, M.A. and Anderson, M.M. 1988. Fruit ripening in *Vitis vinifera* L.: responses to seasonal water deficits. *Am. J. Enol. Vitic.* **39**: 313–20.

May, P. 1965. Reducing inflorescence formation by shading individual sultana buds. *Aust. J. Biol. Sci.* **18**: 463–73.

May, P. 1966. The effect of direction of shoot growth on fruitfulness and yield in sultana vines. *Aust. J. Agric. Res.* **17**: 479–90.

May, P. and Antcliff, A.J. 1963. The effect of shading on fruitfulness and yield in the sultana. *J. Hort. Sci.* **38**: 85–94.

May, P., Clingeliffer, P.R. and Brien, C.J. 1976. Sultana (*Vitis vinifera* L.) canes and their exposure to light. *Vitis* **14**: 278–88.

Muller-Thurgau, H. 1898. Abhängigkeit der Ausbildung der Traubenbeeren und einiger anderer Früchte von der Entwicklung der Samen. *Landw. Jahrb. Schweiz.* **12**: 135–205.

Mullins, M.G.. 1966. Test-plants for investigation of the physiology of flowering in *Vitis vinifera* L. *Nature (Lond.)* **209**: 419–20.

Mullins, M.G. 1967. Morphogenetic effects of roots and of some synthetic cytokinins in *Vitis vinifera* L. *J. Exp. Bot.* **18**: 206–14.

Mullins, M.G. 1968. Regulation of inflorescence growth in cuttings of the grapevine (*Vitis vinifera* L.). *J. Exp. Bot.* **19**: 532–43.

Mullins, M.G. 1980. Regulation of flowering in the grapevine (*Vitis vinifera* L.). In *Plant growth substances 1979* (ed. F. Skoog), pp. 323–30. Springer-Verlag, Berlin.

Muthukrishnan, C.R. and Srinivasan, C. 1974. Correlation between yield quality and petiole nutrients in grapes. *Vitis* **12**: 277–85.

Negi, S.S. and Olmo, H.P. 1966. Sex-conversion in a male *Vitis vinifera* L. by a kinin. *Science* **152**: 1624–5.

Nitsch, J.P. 1953. The physiology of fruit growth. *Ann. Rev. Plant Physiol.* **4**: 199–236.

Nitsch, J. P. and Nitsch, C. 1965. Présence de phytokinines et autres substances de croissance dans la sève d'*Acer saccharum* et de *Vitis vinifera*. *Bull. Soc. Bot. Fr.* **112**: 11–19.

Ough, C.S., Stevens, D. and Almy, J. 1989. Preliminary comments on effects of grape vineyard nitrogen fertilization on subsequent ethyl carbamate formation in wines. *Am. J. Enol. Vitic.* **40**: 219–20.

Ough, C.S., Stevens, D., Sendovski, T., Huang, Z. and An, D. 1990. Factors contributing to urea formation in commercially-fermented wines. *Am. J. Enol. Vitic.* **41**: 68–73.

Palejwala, V.A., Parikh, H.R. and Modi, V.V. 1985. The role of abscisic acid in the ripening of grapes. *Physiol. Plant.* **65**: 498–502.

Peynaud, E. and Ribéreau-Gayon, P. 1971. The grape. In *The biochemistry of fruits and their products* (ed. A.C. Hulme), vol. 2, pp. 171–205. Academic Press, London.

Pirie, A. and Mullins, M.G. 1976. Changes in anthocyanin and phenolics content of grapevine leaf and fruit tissues treated with sucrose, nitrate and abscisic acid. *Plant Physiol.* **58**: 468–72.

Pirie, A. and Mullins, M.G. 1977. Interrelationships of sugars, anthocyanins, total phenols and dry weight in the skin of grape berries during ripening. *Am. J. Enol. Vitic.* **28**: 204–9.

Possingham, J.V. 1970. Aspects of the physiology of grapevines. In *Physiology of tree crops* (ed. L.C. Luckwill and C.V. Cutting), pp. 335–49. Academic Press, London.

Possner, D., Ruffner, H.P. and Rast, D.M. 1983. Regulation of malic acid metabolism in berries of *Vitis vinifera*. *Acta Hort.* **139**: 117–22.

Rives, M. 1972. L'initiation florale chez la vigne. *Conn. Vigne Vin.* **2**: 127–46.

Ruffner, H.P. and Hawker, J.S. 1977. Control of glycolysis in ripening berries of *Vitis vinifera*. *Phytochemistry* **16**: 1171–5.

Ruffner, H.P., Possner, D., Brem, S. and Rast, D.M. 1984. The physiological role of malic enzyme in grape ripening. *Planta* **160**: 444–8.

Sachs, R.M. 1977. Nutrient diversion: An hypothesis to explain the chemical control of flowering. *Hort. Sci.* **12**: 220–2.

Sachs, R.M. and Hackett, W.P. 1976. Chemical control of flowering. *Acta. Hort.* **68**:29–49.

Scienza, A., Miravalle, R., Visai, C. and Fregoni, M. 1978. Relationships between seed number, gibberellin and abscisic acid levels and ripening in Cabernet Sauvignon grape berries. *Vitis* **17**: 361–8.

Shaulis, N.J., Amberg, H. and Crowe, E. 1966. Response of Concord grapes to light exposure and Geneva Double Curtain training. *Proc. Amer. Soc. Hort. Sci.* **89**: 268–80.

Sigiura, A., Utsunomiya, N. and Kobayashi, A. 1975. Effects of daylength and temperature on growth and bunch differentiation of grapevines. *Jap. J. Hort. Sci.* **43**: 387–92.

Singleton, V.L., Salgues, M., Zaya, J. and Trousdale, E. 1985. Caftaric acid disappearance and conversion to products of enzymic oxidation in grape must and wine. *Am. J. Enol. Vitic.* **36**: 50–6.

Singleton, V.L., Zaya, J. and Trousdale, E. 1986. Compositional changes in ripening grapes: caftaric and coumaric acids. *Vitis* **25**: 107–17.

Skene, K.G.M. 1970. The relationship between the effects of CCC on root growth and cytokinin levels in the bleeding sap of *Vitis vinifera* L. *J. Exp. Bot.* **21**: 418–31.

Skene, K.G.M. and Kerridge, G.H. 1967. Effect of root temperature on cytokinin activity in root exudate of *Vitis vinifera* L. *Plant Physiol.* **42**: 1131–9.

Smart, R.E., Turkington, C.R. and Evans, C.J. 1974. Grapevine responses to furrow and trickle irrigation. *Am. J. Enol. Vitic.* **25**: 62–6.

Smart, R.E., Robinson, J.B., Due, G.R. and Brien, C.J. 1985. Canopy microclimate modification for the cultivar Shiraz. II. Effects on must and wine composition. *Vitis* **24**: 119–28.

Srinivasan, C. and Mullins, M.G. 1976. Reproductive anatomy of the grapevine (*Vitis vinifera* L.): Origin and development of the Anlage and its derivatives. *Ann. Bot.* **38**: 1079–84.

Srinivasan, C. and Mullins, M.G. 1978. Control of flowering in the grapevine (*Vitis vinifera* L.): Formation of inflorescence in vitro by isolated tendrils. *Plant Physiol.* **61**:127–30.

Srinivasan, C. and Mullins, M.G. 1979. Flowering in *Vitis*: Conversion of tendrils into inflorescence and bunches of grapes. *Planta* **145**: 187–92.

Srinivasan, C. and Mullins, M.G. 1980a. Effects of temperature and growth regulators on formation of Anlagen, tendrils and inflorescence in *Vitis vinifera* L. *Ann. Bot.* **45**: 439–46.

Srinivasan, C. and Mullins, M.G. 1980b. Flowering in the grapevine (*Vitis vinifera* L.): Histochemical changes in apices during the formation of the Anlage and its derivatives. *Zeitsch. Pflanzenphysiol.* **97**: 299–308.

Srinivasan, C. and Mullins, M.G. 1981*a*. Physiology of flowering in the grapevine – a review. *Am. J. Enol. Vitic.* **32**: 47–63.

Srinivasan, C. and Mullins, M.G. 1981*b*. Induction of precocious flowering in grapevine seedlings by growth regulators. *Agronomie* **1**: 1–5.

Srinivasan, C., Muthukrishnan, C.R. and Shivashankara, K.T. 1972. Influence of nutrients on the size of cluster primordia in grape buds (*Vitis vinifera* L.). *Potash Review* **29**: 1–4.

Steffan, H. and Rapp, A. 1979. Ein Beitrag zum Nachweis unterschiedlicher Malatpools in Beeren der Rebe. *Vitis* **18**: 100–5.

Takimoto, K, Saito, K. and Kasai, Z. 1976. Diurnal change of tartrate dissimilation during the ripening of grapes. *Phytochemistry* **15**: 927–30.

Tse, A.T.Y., Ramina, A., Hackett, W.P. and Sachs, R.M. 1974. Enhanced inflorescence development in Bougainvillea 'San Diego Red' by removal of young leaves and cytokinin treatments. *Plant Physiol.* **54**: 404–7.

Weaver, R.J., Van Overbeer, J. and Pool, R.M. 1965. Induction of fruit set in *Vitis vinifera* L. by a kinin. *Nature (Lond.)* **206**: 952–3.

Webb, A.D. 1981. Quality factors in California grapes. In *Quality of selected fruits and vegetables of North America* (ed. R. Teranishi & H. Barrera-Benitez) (American Chemical Society Symposium Series no. 170), p. 9.

Winkler, A.J., Cook, J.A., Kliewer, W.M. and Lider, L.A. 1974. *General viticulture.* Second edition. University of California Press, Berkeley. 710 pp.

Zeevaart, J.A.D. 1976. Physiology of flower formation. *Ann. Rev. Plant Physiol.* **27**: 321–48.

6

The cultivated grapevine

Radiation and energy transfer at the vine's surface

The sun is the most important source of radiant energy; this energy is emitted in a continuous spectrum. Radiation of wavelengths between 150 and 3000 nm is termed shortwave or solar radiation. Radiation with wavelengths from 3×10^3 to 100×10^3 nm is known as longwave radiation. The atmosphere prevents approximately half of the shortwave radiation from reaching the Earth's surface. The atmosphere is a relatively poor absorber of shortwave radiation but it is a good absorber of longwave radiation. This is due to the absorption of longwave radiation by water vapor, carbon dioxide and ozone. This absorption of longwave radiation by the atmosphere contributes to a net warming of the Earth's surface. The recent concern over the rapid increase in the CO_2 partial pressure of the atmosphere, as a consequence of increased fossil fuel consumption and deforestation, is founded on the prediction that there will be a rapid warming of the Earth ("greenhouse effect") because less longwave radiation will be lost to space.

Radiant energies between 300 and 1000 nm, the so called 'biological window', have a major influence on life processes (Hart, 1988). The energy of radiation is inversely proportional to its wavelength. Therefore, radiation of wavelengths greater than 1000 nm has very little energy, but the energy of wavelengths less than 300 nm is sufficient to be destructive to living organisms.

The light reaching the earth's surface is able to induce various biological responses through changes in its quality, quantity, direction and periodicity. For plants, these responses include thermal effects, photomorphogenesis, photosynthesis and mutagenesis.

The First Law of Thermodynamics states that energy can be neither created nor destroyed but can be changed from one form to another. When radiation strikes a plant's surface the energy flux into and out of

147

an organ must therefore equal the rate of energy storage in that organ. The energy absorbed by a leaf may be in the form of shortwave (from the sun or sky) or longwave radiation. The flux of energy out of a leaf may be emitted infrared radiation (IR), sensible heat (heat loss through conduction and convection), and latent heat (heat loss by transpiration). The storage of energy in the leaf is in the form of chemical bond energy (products of photosynthesis) and physical storage (heat capacity of the organ). Photosynthesis of plant leaves stores less than one per cent of the absorbed radiation and the physical storage is small unless the organ is quite large. Therefore, without the cooling processes of IR emittance and sensible and latent heat loss, the temperature of grapevine leaves, or other organs under full sunlight, would increase very rapidly.

The loss of heat by conduction is analogous to diffusion (transfer of heat along a temperature gradient). Convection of heat is the transfer of energy due to the movement of small 'packets' of air. The movement of energy at the plant's surface is influenced by its boundary layer. The boundary layer of an organ or canopy is the unstirred layer of slowed molecular motion adjacent to its surface. This arises at a solid–fluid interface owing to friction and fluid viscosity. The boundary layer thickness is influenced by wind speed, path length and surface topography. The boundary layer thickness for a grape leaf is approximately 4.0 $(l/v)^{1/2}$, and that for a berry is 2.8 $(d/v)^{1/2} + 0.25/v$, where l = path length, v = wind velocity and d = diameter. The boundary layer is always an important factor for loss of sensible heat from leaves. For latent heat loss, the boundary layer is significant only if stomata are open.

The temperature of a leaf is dependent upon various environmental conditions and conductances to water vapor flux within the lamina and at the boundary layer. Whether the latent heat loss is sufficient to cool the leaf below ambient temperature is dependent upon (i) the radiation heat load and (ii) the concentration gradient of water vapor from inside the leaf to the atmosphere. Any increase in the vapor pressure difference between the leaf and atmosphere generally increases latent heat loss provided that other conductances within the soil–plant–air continuum are non-limiting. Increasing the radiative heat load on the leaf while maintaining other factors constant always tends to increase leaf temperature.

The temperature of a berry is probably the most important environmental factor during its development and ripening. Once berries have reached veraison, the few lenticels on their surface contribute little to the diffusive conductance (i.e. there is only a small loss of latent heat). Heat loss by re-radiation of longwave radiation also is very small for berries. The effect of wind speed on berry surface temperature is very important because most cooling of the clusters occurs by forced convection. Therefore, the temperature of the fruit of grapevines is dependent upon the radiation load at the surface of the clusters. Cultural practices are important in determining the relative exposure of bunches to solar radiation and have a major effect on the thermal environment of the fruit.

The microclimate around and within the vine's canopy is influenced by the number of shoots that emerge per unit row length and by the length to which the shoots grow (this ultimately determines the vine's total leaf area). Wind speed, evaporative demand, photon flux density and the red : far-red ratio are the environmental parameters which are altered by the vine's canopy. Humidity and air temperature are less affected by the canopy and are likely to be similar to the ambient environment. A generalized scheme of the different environmental parameters that influence the growth of the vine and their effects on the physiology of the vegetative and reproductive organs is shown in Fig. 6.1.

Effects of cultural practices on light interception, growth and yield

SOLAR RADIATION INTERCEPTION AND VINE MICROCLIMATE

Plants may absorb, reflect or transmit light that strikes their surfaces. The unique feature of leaves is their ability to absorb strongly in the 400–700 nm portion of the solar radiation spectrum. Plant physiologists term these wavelengths 'photosynthetically active radiation' (PAR). As mentioned in Chapter 4, a single grape leaf will absorb 80–90% of the PAR that strikes it. Once the vine's canopy is composed of several leaf layers, leaves to the interior will rarely be above the light saturation point because the intensity of the transmitted light will be only 10–20% of that of normal sunlight.

The amount of direct light absorbed by the vine's top and by the walls of the canopy is the dominant component of solar radiation involved in photosynthesis (Smart, 1973). When there are no clouds, diffuse light interception contributes a smaller fraction of the total carbon gain of the vine under such conditions. On cloudy days the importance of diffuse light for photosynthesis increases considerably. It has been estimated that direct solar radiation intercepted by the leaves which comprise the outer leaf layer accounts for approximately 70% of the carbon fixed once the vine is at full canopy (Smart, 1974). The overriding impor-

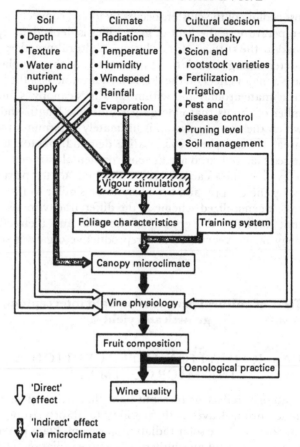

Fig. 6.1. Relationships among environmental, edaphic and cultural factors, their influence on the growth of grapevines, and their resulting effects on fruit composition and wine quality. From Smart (1985). Reproduced with permission

tance of the outer leaf layers of the canopy in grapevine photosynthesis was demonstrated by Williams *et al.* (1987). The removal of the interior portion of the vine's canopy (approximately 30% of the vine's leaf area) did not delay berry growth or maturation of the fruit on the vines over a two-year period.

The remaining 30% of the CO_2 fixed by the vine is by leaves exposed to diffuse light or sunflecks. The leaf area of the canopy illuminated by diffuse shortwave radiation is greater than that illuminated by direct radiation. However, diffuse light is of low intensity and photosynthesis by these leaves is minimal. Sunflecks occur when direct solar radiation enters the gaps present in the canopy. The significance of sunflecks in grapevine photosynthesis is discussed by Kriedemann *et al.* (1973).

Another important aspect to consider with regard to radiation within the canopy is the quality of light that reaches the interior portion, as many aspects of plant morphogenesis are mediated by the quality of light. The photoreceptors within the plant which perceive light quality are phytochrome, cryptochrome (a blue-light-absorbing receptor) and chlorophyll. The phytochrome system within plant organs is regulated by the ratio of the red wavelength (maximum absorption at 660 nm) to far-red wavelength (maximum absorption at 730 nm) (R : FR). The R : FR ratio decreases once solar radiation passes through a leaf because light within the red wavelengths is absorbed to a greater extent than that within the far-red wavelengths. These differences have significant effects on the growth, differentiation and metabolism of organs.

EFFECTS OF ROW DIRECTION AND VINE DENSITY (NUMBER PER UNIT AREA)

Row direction within the vineyard influences the interception of solar radiation by vines (Smart, 1973), and this is affected by the stage of vine development and trellis type. At full canopy development the top and sides intercept most of the direct solar radiation. Radiation interception by the vine after full canopy development is considerably more efficient for vines planted in north–south rows than those planted in an east–west direction. The quantum flux densities on the sides and tops of vines planted in both north–south and east–west directions are illustrated in Fig. 6.2. The advantage of the north–south direction is that the east and west sides of the vine's canopy are exposed to direct solar radiation in both the morning and the afternoon. Therefore, the leaves on these sides are at light saturation for photosynthesis during the period of the day in which the sun is shining directly on them. Conversely, the leaves on the north and south sides of vines planted in east–west rows may not reach full light saturation at any time during the day, even during the longest days of the year in California (Fig. 6.2). In both row directions the leaves at the top of the canopy are always able to receive the maximum amount of solar radiation that is available throughout the day.

Changing the vine density within the vineyard also affects the total amount of solar radiation intercepted per unit ground area (Smart, 1973). Many vineyards in Europe have very high vine densities (1 × 1 m spacings), but the individual vines are short in stature. The total radiation intercepted by these small vines may be less than that of a taller vine

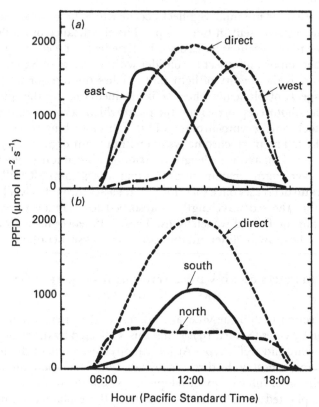

Fig. 6.2. The photosynthetic photon flux density (PPFD) at the canopy surface of grapevines planted to either north–south (*a*) or east–west (*b*) rows in the San Joaquin Valley of California. To obtain the direct reading a single quantum sensor was placed at the top of the canopy perpendicular to the soil surface. Line quantum sensors (1 m in length) were placed at the approximate angle of the canopy for readings taken on either side of the vine. Measurements were taken in the month of June

spaced at 2 or 3 m. The total amount of leaf area per unit land area for Chenin blanc vines planted at row spacings of 3.66 and 2.44 m were approximately 28 200 and 27 500 m²ha⁻¹, respectively (Table 6.1). However, when the surface area of the canopy was measured, the value of leaf area per unit land area was almost 37% greater for the closer-spaced vines than those spaced at 3.66 m. These results indicate that the closer the vines are planted the more efficient they are in intercepting solar radiation, owing to greater leaf surface per unit land area even though total leaf area per vine is less.

Table 6.1. *The effects of vine density and trellis system on leaf area and the partitioning of dry matter among vegetative and reproductive growth of Chenin blanc grapevines*

Vine density[a] (ha^{-1})	Trellis type[b] (m)	(no.)	Leaf area (m² ha^{-1})	Canopy surface area (m² ha^{-1})	Dry mass (g vine^{-1}) shoot	cordons	trunk	roots	Yield vine^{-1} (kg)	ha^{-1} (t)
1121	1.07	(2)	28 168	11 344	5622	3978	2881	2689	27.6	30.9
1680	1.07	(2)	27 451	15 490	3909	3469	2845	2669	24.1	40.5
1680	1.62	(0)	34 552	12 751	4271	3581	2391	2500	26.4	44.4
1121	1.62	(0)	—[c]	11 098	4271	—	2391	—	28.3	31.7

[a] Row and vine spacings were 3.66 m × 2.44 m, to achieve a vine density of 1121 ha^{-1}, and 2.44 m × 2.44 m, to achieve a vine density of 1680 ha^{-1}. For further information on the culture of these vines, see Table 4.1.

[b] The first number in this column refers to the height (m) of the cordon wire. The second number indicates the number (no.) of foliage wires above the cordon.

[c] Data not collected.

TRELLIS SYSTEM

The trellis system plays a major role in determining the interception of solar radiation by the vine (Fig. 6.3). One means of overcoming the low solar radiation intensity of an east–west row is to spread the canopy so that more leaves are exposed to solar radiation throughout the day. This also reduces the amount of soil exposure between the vine rows. The interception of solar radiation of a north–south vine row can be increased by spreading the canopy or by increasing the height of the trellis system. Increasing the height of the canopy allows more foliage on either side of the row to absorb direct solar radiation for a longer period of time during the day.

The trellis system also affects the microclimate within the vine's canopy. The microclimate near and inside the canopy differs from the ambient conditions directly above the vineyard according to the number and arrangement of the vine's leaves and the volume allocated for the canopy. The number of leaves is a function of how the vine is trained and the pruning pattern. The arrangement and volume of the canopy is influenced by the trellis system.

In a review by Smart (1985) several different trellis systems were compared to demonstrate their effects on the vine's canopy. The different trellis systems changed the arrangement of the leaves within the canopy and the direction of shoot growth. It was shown that both leaf area and canopy surface area per unit land area varied considerably with the trellis system. The canopy surface area was greatest when the canopy was separated. The Geneva Double Curtain (GDC) system had more than twice the surface area of the Tendone overhead arbor system. These systems also differed with regard to the vine's interior microclimate. Splitting the vine's canopy, or increasing vine density, resulted in differences in the amount of light intercepted and in the amount of shading in the fruit zone.

VINEYARD IRRIGATION

Vineyard irrigation determines the vigor of vine growth and affects the canopy size and microclimate. The application of excess water is wasteful and may favor vegetative growth. Conversely, irrigation practices that result in soil water deficits will adversely affect the physiological functions of the grapevine, including photosynthesis. The most important effect of a soil water deficit on whole vine photosynthesis is that of reducing leaf area necessary for efficient solar radiation interception. The reduction in leaf area by soil water deficits is due either to inhibition

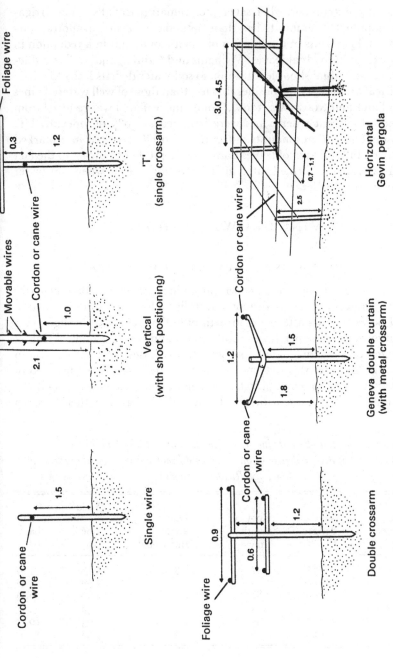

Fig. 6.3. Diagrammatic representation of six trellis systems. The use of a particular trellis is dependent upon numerous factors to include cultivar, intended use of the grapes and climatic and edaphic conditions. The dimensions (in meters) are approximate values

of shoot elongation or, when severe, to premature leaf abscission. Irrigation supplied in amounts less than potential vineyard evapotranspiration (ET_c) results in a reduction of shoot length and in a reduction in the area of individual leaves (Williams and Grimes, 1987). The reduction in individual shoot lengths due to soil water deficits leads to whole vine leaf areas that are considerably less than those of well watered vines (Williams and Matthews, 1990). Total vine leaf area was reduced from 35 to 50 percent when vines were irrigated at 40% of vineyard ET. Seasonal irrigation amounts of less than 1.0 ET_c resulted in a marked decrease in total area of leaves borne on lateral shoots, owing to the fact that laterals are formed later in the growing season when soil water deficits are prevalent.

VINE GROWTH AND YIELD

VINE DENSITY (NUMBER PER UNIT AREA)

The planting density of vines within a vineyard affects the vineyard's growth and productivity throughout its life. The spacing of vines within and between rows affects the growth of individual vines through competition with their neighbors, both above and below ground. While it is commonly assumed that root growth decreases with an increase in planting density, this may not always be the case (Table 6.1). Root density in the top 0.6 m of soil (root length per cubic meter of soil) has been shown to increase with vine density (Table 6.2). Whether there is a

Table 6.2. *The effect of vine density on the total root length in the top 0.6 m of soil, vine leaf area and yield of three-year-old Pinot noir grapevines grafted onto 99 Richter rootstock*

Vine spacing (m)	Vine density (ha^{-1})	Root length (m vine^{-1})	Leaf area (m^2 vine^{-1})	Yield	
				vine^{-1} (kg)	ha^{-1} (t)
3.0 × 3.0	1111	5.97	6.3	4.12	4.57
3.0 × 1.5	2222	4.68	4.5	2.50	5.51
2.0 × 2.0	2500	4.84	4.0	2.60	6.54
2.0 × 1.0	5000	4.13	4.0	1.43	5.15
1.0 × 1.0	10000	2.93	2.7	1.03	10.30
1.0 × 0.5	20000	2.46	1.3	0.58	11.60

Adapted from Archer (1987) and Archer and Strauss (1985).

reduction in root biomass as vine density increases is dependent upon available soil rooting depth, soil water availability and other edaphic factors.

Vine density affects the above-ground vegetative growth. Winkler (1969) found that as the vine density of both Cabernet Sauvignon and White Riesling in an unirrigated vineyard decreased from 2000 to 667 vines ha^{-1} the trunk circumference increased. The amount of shoot growth and leaf area per vine both decrease with increasing vine density (Tables 6.1 and 6.2), but this may or may not increase the total leaf area per unit land area. The reduction in shoot growth per vine with closer spacing also is reflected in lower pruning weights per vine (Brar and Bindra, 1986; Lavee and Haskal, 1982; Shaulis and Kimball, 1955). The reduction in shoot growth per vine may be due to the fact that the closer the vines are spaced the more efficient they become at utilizing soil water and the more rapidly they deplete the available water in the profile (Archer et al., 1988).

As vine density increases, yield per vine decreases but yield per unit land area generally increases (Brar and Bindra, 1986; Hedberg and Raison, 1982; Hunter et al., 1985; Lavee and Haskal, 1982; Shaulis and Kimball, 1955) (Tables 6.1 and 6.2). There are a few reports that yield per unit land area does not increase with an increase in vine density (Winkler, 1969) and that yield is dependent upon the arrangement of the vine and row spacings (Shaulis et al., 1966). Generally, the composition of the fruit (soluble solids, titratable acidity and pH) is not significantly affected by the increased yields at the higher vine densities (Hunter et al., 1985; Shaulis et al., 1966).

TRELLIS TYPE

The trellis system used in the vineyard greatly affects the amount of vegetative growth and yield of the vines. Trellis systems that promote vigorous shoot growth and high yields also promote high root numbers and density in the soil (Archer et al., 1988; Van Zyl and Van Huyssteen, 1980). The percentage of the total number of roots of different sizes (based upon their diameters) in the soil profile is not affected by the trellis system nor is their distribution. The greatest effect of trellis size on the number of roots is in the occurrence of fine roots. It should be noted that fine roots do not contribute much to the total root biomass (McKenry, 1984; Williams and Smith, 1991); the contention that vigorous above-ground growth is balanced by a larger root system (Archer et al., 1988) may not always be true.

Effects of trellis type on the growth of the above-ground permanent structures (trunk and cordons) of the vine have received little attention. Increasing the height of the trellis, and thereby increasing the length of the trunk, resulted in a reduction in trunk dry mass in Chenin blanc grapevines (Table 6.1). Changing the height of the trellis did not have an effect on the weight of the cordons in vines trained to bilateral cordons. Little is known of the partitioning of dry weight to the cordons when the training system includes quadrilateral cordons. According to Saayman and Van Huyssteen (1980) the mass of the trunk and cordons of vines trained with the same type of trellis system was dependent upon the amount of growth of the shoots over a ten-year period.

Increasing the height of the trellis generally increases the amount of vegetative growth and vine yield (Shaulis and Robinson, 1953; Shaulis et al., 1953; Weaver et al., 1984) (Table 6.1). This increase of yield of vines growing on tall trellises is due to an increase in the number of clusters that develop from the retained buds, and increase in berry mass. The use of crossarms to divide the canopy of the grapevine may also lead to an increase in the productivity of the vine (Shaulis and May, 1971; Shaulis et al., 1966). Again, the increase in yield is due to a greater number of clusters per vine. Increasing the height of the vine, or splitting the vine's canopy by the use of a trellis, provides good exposure of next year's fruiting canes or spurs to solar radiation (Kliewer, 1982). The positive effect of solar radiation on the fruitfulness of grape buds has been demonstrated by several researchers (Baldwin, 1964; May, 1965; May and Antcliff, 1963).

The trellis and training systems greatly affect the microclimate within the vine's canopy, and they influence the amount and distribution of leaves in space, which, in turn, affects the interior microclimate. Differences attributed to the effects of trellis systems on the quality of the fruit at harvest may be due to the effects of microclimate on fruit composition (Carbonneau et al., 1978; Smart, 1985; Smart et al., 1985a,b). The trellis system together with leaf area per unit row length determines the amount of light that reaches the cluster zone in the canopy. Trellis and training systems that cause excessive shading of clusters may produce fruit of inferior quality. Shading causes increases in the potassium concentration, pH, and malic acid content of berries and a decrease in berry size, soluble solids, phenols, anthocyanins and monoterpenes (Crippen and Morrison, 1986; Dokoozlian, 1990; Reynolds and Wardle, 1989; Smart et al., 1985b). Differences in the composition of fruit exposed to solar radiation may be due to thermal effects and/or to effects of light quantity or quality. Kliewer and Smart (1989) concluded that phytochrome was involved in changing the composition of the fruit when it was shaded, but Dokoozlian (1990) found that light quality (R : FR ratio) had no influence on the growth or compositional development of the fruit; in the latter study changes in fruit composition were found to be due to effects of light quantity.

The trellis system also affects the incidence of pathogens within the vineyard (Savage and Sall, 1984). The major effect of a trellis system on the control of *Botrytis* is correlated with the ventilation of the fruiting zone (Savage and Sall, 1984). Another cultural practice used in conjunction with trellising to improve canopy microclimate is the removal of basal leaves and lateral shoots from the primany shoots (Bledsoe *et al.*, 1988). The removal of these leaves also lessens the incidence of *Botrytis* (Gubler *et al.*, 1987). The removal of basal leaves increases the movement of air into and out of the cluster zone and this causes an increase in evaporation and drying of the foliage, factors which discourage fungal activity (Thomas *et al.*, 1988).

The choice of trellis system for a particular vineyard is determined by the end use of the grapes, the climatic conditions of the area, soil characteristics and pruning and harvesting methods. Excessive exposure of clusters in a hot growing region will reduce the quality of the fruit because of ambering (discoloration) and desiccation of the berries, particularly in white wine or table grape cultivars. The production of red table grapes requires diffusion of sufficient light into the fruiting zone to aid in the coloring of the berries while keeping direct solar radiation to a minimum. Vineyards that are mechanically pruned and/or harvested require trellis systems that allow the machinery to perform at maximum efficiency. This means that the trellis system must be designed to fit the specific piece of equipment to be used and to withstand the stresses that the machinery may place on the trellis during its operation.

PRUNING LEVEL

Pruning of grapevines is designed to establish and maintain the vine in a specific form, to save labor and facilitate vineyard operations. The objective of pruning is to distribute the bearing wood over the vine and to lessen the need for thinning as a means of controlling the crop load (Winkler *et al.*, 1974). Overcropping can be deleterious to vines of certain cultivars under specific conditions (Weaver and McCune, 1960; Winkler, 1929, 1954). The use of either spur pruning (short, 2–3-node bearing wood) or cane pruning (bearing wood of 10–15 nodes) is determined by whether the basal nodes on the cane are fruitful. This is a characteristic of the cultivar. Traditional spur pruning removes more than 70% of the potential bearing wood in any one year. More recent work has indicated that the use of minimal pruning (the retention of significantly more buds per vine than traditional pruning) can increase crop yield without affecting fruit quality and production in subsequent years (Clingeleffer, 1984; Jackson et al., 1984; Reynolds, 1988). Minimal pruning involves the use of tractor-mounted saws, which trim or 'skirt' the canopy on either side of the row. The crown or top of the vine is usually left unpruned. Minimal pruning is effective in maintaining yield and quality provided that the trellis system provides sufficient space for efficient display of the foliage. The use of wide crossarms or split canopies are essential components of minimal pruning systems. The use of minimal pruning of vines is being investigated in most major grape-growing regions of the world.

The optimal production of high-quality grapes in a given vineyard is dependent upon the selection of the appropriate pruning level, the training or trellis system and end use of the crop. The production of table grapes does not lend itself to minimal pruning. The reduction in cluster and berry size, which accompanies an increase in the numbers of buds retained, results in inferior fruit for the table market.

Mineral nutrition

USE OF MINERAL NUTRIENTS BY GRAPEVINES

This discussion will be limited to the three macronutrients and four micronutrients that generally limit grape production in world viticulture. Other macro- and micronutrients may be important at the local level. Mineral nutrients are used by grapevines for various physiological processes and as structural components (Table 6.3). Nitrogen (N) concentrations in grapevine tissues are highest in the leaves and are next highest in the root system. The concentrations in the smaller, feeder, roots are greater than in the larger structural portions of the root. The high concentration of N in the leaves is expected because they contain the enzymes for the PCR cycle, for carbohydrate metabolism, and for nitrate reduction and metabolism. RuBPC/O alone may constitute up to 50% of the total protein within the chloroplast. Nitrogen is also a structural component of the chlorophyll molecule. Thus, a deficiency of N in grapevines results in the chlorosis of the leaves. Chlorosis occurs on the basal leaves first because N can be translocated from the older tissues to the younger ones. The foliar deficiency symptoms of other mineral nutrients may appear first on the apical or basal leaves depending upon their mobility within the vine.

Vineyards that become deficient in N usually exhibit a gradual de-

Table 6.3. *Functions and deficiency symptoms of common macro- and micronutrients*

Mineral nutrient	Major functions	Deficiency symptoms
Nitrogen	Structural component of proteins, nucleic acids and chlorophyll	Chlorosis of basal leaves; yield reduction
Potassium	Involved in carbohydrate metabolism and transport; stomatal functioning; acts as an osmoticum	Chlorosis of basal leaves; yield reduction
Phosphorus	Used in high-energy bonds	Chlorosis of leaves to include reddening on leaves of black or red cultivars; reduced berry set
Magnesium	Found in chlorophyll molecule; enzyme activator	Chlorosis of basal leaves
Zinc	Involved in synthesis of indoleacetic acid; involved with pollination, chloroplast development	Chlorosis of apical leaves first; formation of shot berries
Manganese	Involved in the splitting of water in light reactions of photosynthesis; enzyme activator	Chlorosis of basal leaves
Boron	Involved with carbohydrate metabolism, pollination and fertilization	Chlorosis of apical leaves first; formation of shot berries
Iron	Involved in chlorophyll biosynthesis; enzyme activator	Chlorosis of apical leaves first

crease in crop yield before the occurrence of foliar deficiency. This is due to an overall reduction in vine growth by lack of N; the application of N fertilizers will overcome this problem. The over-fertilization of vineyards with N causes an excess of vegetative growth and a reduction in yield. This may be due to the creation of a very dense canopy, which causes a reduction in the light that reaches the buds that will be retained for the next year's fruiting wood. These shaded buds have a lowered bud fruitfulness. The continued use of N fertilizers in a non-deficient situation causes an increase in vegetative growth rather than an increase in the N concentration of the vine's organs.

Large amounts of potassium (K) are required by grapevines, but K is not a part of organic molecules. Potassium is readily translocated within the vine and it may be involved in the translocation of carbohydrates. Potassium is used as an osmotic agent in the opening and closing of stomata (Raschke, 1979). Potassium neutralizes organic acids and it plays an important role in controlling the acidity and pH of the fruit's juice. It has been hypothesized that hydrogen ions may be exchanged with K ions, thereby increasing the juice pH (Boulton, 1980a,b). Cultural practices may affect the partitioning of K within the vine and its distribution and/or remobilization from other organs to the fruit (Smart et al., 1985a, b). The reduction in yields due to K deficiencies is due to the reduction in the vegetative growth of the vine.

Phosphorus is another macronutrient that is mobile within the vine. It is a part of nucleoproteins and phospholipids. Phosphate groups are involved in energy transfer processes. A deficiency of phosphorus results in a reduction of vine leaf photosynthesis. Severe deficiencies of phosphorus cause a reduction in the set and yield of grape berries (Skinner and Matthews, 1989).

Magnesium (Mg) is a part of the chlorophyll molecule, and its deficiency causes chlorosis of the foliage. Mg is also a mobile element, so deficiency symptoms first appear in the basal leaves of the shoot. Magnesium is involved in the activation of many enzymes and in maintaining the integrity of ribosomes.

Deficiencies of several micronutrients are major problems in vineyards worldwide. A deficiency of zinc (Zn) in grapevines results in the formation of leaves that are smaller than normal and/or mottled, and shortened internodes. This reduction in shoot growth may result from the fact that Zn is essential for the synthesis of tryptophan, a precursor of the phytohormone indoleacetic acid (IAA). Foliar symptoms of zinc deficiency also include interveinal chlorosis. Zinc is tightly bound to many enzymes and is essential for their functioning. Another Zn deficiency symptom in grapevines is the production of clusters with few berries; these berries may vary in size from normal to very small.

Another micronutrient deficiency in grapevines that results in clusters with berries of variable size is boron (B). Boron is involved in the elongation of pollen tubes, and it is needed for proper cell division in the root and shoot apices and young leaves. Boron is not readily translocated within the plant. Boron appears to be needed for sugar translocation. In cases of B deficiencies, shoots may have a zigzag appearance. The foliar symptoms of B deficiency are easily confused with other grapevine maladies. A unique feature of B in plants is that the concentration range between deficiency, sufficiency and toxicity is very narrow. Grapevines are sensitive to an excess of B in the soil.

The foliar deficiency symptoms of both manganese (Mn) and iron (Fe) are chlorosis of the leaves. In grapevines, Mn deficiency symptoms occur first on the basal leaves whereas those of Fe generally occur on the apical leaves. Manganese is part of the structure of the chlorophyll molecule and an activator of numerous enzyme systems. Manganese also has a role in the splitting of water, which occurs during the light reactions of photosynthesis. Iron is involved in the synthesis of chlorophyll and is part of certain enzyme systems. It undergoes oxidation and reduction during the electron transport component of photosynthesis and respiration.

ASSESSMENT OF NUTRIENT DEFICIENCIES OR TOXICITIES

The presence of visual symptoms generally indicates that the deficiency has already had a deleterious effect on vine growth or yield. A gradual reduction in yield over the years may indicate a need for a specific nutrient but the vines concerned may exhibit no recognizable symptoms. Some form of analysis is required to establish which mineral nutrient is the cause of the yield reduction. Soil analysis can be used to assess problems related to soil pH, salinity or other toxicities, but it is not a reliable means of determining the nutritional requirements of vineyards (Christensen *et al.*, 1978).

A more reliable means to assess nutrient requirements of vineyards is the use of plant tissue analysis. The analysis of the plant tissues provides information on the nutrient ions available to the plant in the soil profile. Frequently, the amount of a mineral element within the soil is in excess but its availability is dependent upon factors such as soil pH, organic matter content and the presence of other ions. By sampling the vine it is possible to obtain a comprehensive picture of the different factors that determine the soil's potential to supply mineral nutrients.

Much effort has gone into establishing the procedures to assess the vine's mineral status by tissue analysis. At present, petioles are the tissue of choice for analysis of vine mineral status in California and elsewhere. In a long-term fertilizer study in South Africa, petiole analyses were the most sensitive indicator of N, P, and K status in Chenin blanc grapevines (Conradie and Saayman, 1989). The procedures for petiole analysis are as follows: leaves are harvested at nodes opposite to the cluster at anthesis or véraison and the blade is removed. The petioles are then washed in distilled water, dried and analyzed for nutrients. Analytical work is usually done by a commercial laboratory on a contract basis. The position on the shoot at which the petioles are chosen, and the stage of vine growth at which petioles are harvested, are important factors in reducing variability from year to year. To increase the accuracy of the sampling procedure, the petioles should be taken at random from vines within the vineyard. Vines from the outside rows and vines at the ends of the rows should be avoided. If the vineyard has several soil types within its borders, samples should be collected from vines in areas with different soil types.

The optimum, deficit and toxic concentrations of the various minerals in petiole samples have been determined for the cultivar Thompson Seedless growing in California (Christensen *et al.*, 1978), but these values may not be applicable to other cultivars or to other regions.

SALINITY

Sodium is not required by most plants, including grapevines, but chloride is required in small amounts for photosynthesis. Both Na and Cl are frequently present in excess; this leads to water stress, a decreased soil permeability and specific ion toxicities. In contrast to most herbaceous crops, grapes are affected more by specific ion toxicities, specifically Cl^-, than by the lowered soil water potential of a saline soil solution.

Salinity problems develop most often in arid and semi-arid regions where evaporation of soil water exceeds percolation. Salts are carried towards the soil surface and concentrate as water evaporates. Grapes have been classified as moderately sensitive to salinity; they are less tolerant than cotton but more tolerant than lettuce. Yield decreases of approximately 10% per dSW m^{-1} are observed (Maas, 1987).

Genetic variability for salt uptake and tolerance is present among *Vitis* species, cultivars and rootstocks. Salt-tolerant cultivars include French Colombard (most tolerant) > Grenache, Chenin blanc, Thompson Seedless > Barbera, Muscat of Alexandria, Ribier (susceptible). *V. vinifera* takes up more Cl than most other species; chloride-excluding rootstocks offer some protection. Rootstocks offer less protection against Na^+ uptake. Rootstocks that have shown some salt tolerance include Dogridge (*V. champinii*), Salt Creek (*V. champinii*), Harmony (*champinii* × 1613), and Schwarzmann. As for most fruit crops, rootstocks exert a major influence on ion uptake. For example, the rootstocks Salt Creek and Dogridge decreased Cl^- uptake three to six times as compared with own-rooted Thompson Seedless (Sauer, 1968). The salt tolerance of species in descending order is: *rupestris* < *berlandieri, riparia* < *candicans, champinii, longii* < *cinerea, cordifolia* < *vinifera* (Downton, 1977).

Severe saline conditions inhibit photosynthesis and delay fruit ripening, but moderate levels (less than 50 m*M* NaCl) have been shown to accelerate ripening (Downton and Loveys, 1978; Hawker and Walker, 1978).

VINE REQUIREMENTS FOR MINERAL NUTRIENTS

The concentrations of most mineral nutrients within the vine are highest early in the season and then decrease as growth continues. An example of the decreases in the concentration of mineral nutrients in the leaves, stems (main axis of the shoot), and clusters of grapevines is shown in Fig. 6.4. The decreases in the concentrations of nitrogen shown here are due to a dilution effect because the total amount of N increased or remained the same as the organs continued to grow. This dilution occurred because accumulation of sugars (as in the clusters) or cell wall components (as in the leaves and stems) increased to a greater extent than the uptake and use of the mineral nutrient. The concentrations of K and P in the leaves also decrease throughout the growing season (Christensen, 1969; Conradie, 1981b). The concentrations of Ca and Mg in the leaves increase or remain the same during vine growth (Conradie, 1981b).

The amounts of mineral nutrients required by grapevines are considerably less than those required by many of the major agricultural crops (Olson and Kurtz, 1982). The amount of N needed for the growth of the current year's shoots and fruit of Thompson Seedless in California (Table 6.4) and in grapevines of the same cultivar growing in Australia was 84 kg ha^{-1} (Alexander, 1958; Williams, 1987). Lafon et al. (1965) determined that 68 kg N ha^{-1} was needed for growth of the shoots and fruit of St Emilion vines; Lohnertz (1988) found that 73 kg N ha^{-1} was needed by Riesling vines. The amount of nitrogen per ton of fruit removed from the vineyard for several different grape cultivars is shown in Table 6.5. The absolute amount of nitrogen in the fruit varies with the cultivar, soil conditions, vineyard location and fertilizer regime.

The amounts of K and P in the vegetative and reproductive structures of the vine are also less than those found in several annual row crops. The greatest requirement for K in grapevines is that of fruit (Table 6.4). According to Williams et al. (1987) and Lohnertz (1988) there is approximately 56 kg K ha^{-1} in the fruit of the vine at harvest. This represents about 60% of the total K found in the current season's above-ground growth; similar results have been reported by Lafon et al. (1965) and Smart et al. (1985a). The amount of P in the current season's growth of field-grown grapevines was about 10 kg ha^{-1} (Lafon et al., 1965). The amounts of other macro- and micro-nutrients

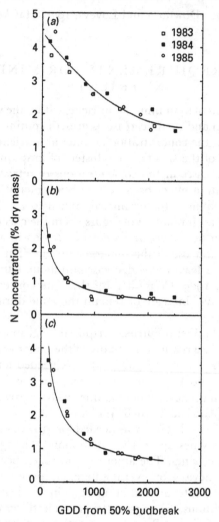

Fig. 6.4. Nitrogen concentrations in the leaves (*a*), stems (main axis of the shoot) (*b*) and clusters (*c*) of Thompson Seedless grapevines as a function of growing degree days (GDD) greater than 10 °C after budbreak. From Williams (1987)

which occur in the grapevine are given by Conradie (1981*a*), Lafon *et al.* (1965), and Lohnertz (1988).

The amounts of mineral nutrients in the permanent structures of mature vines are considerable. The amounts of N and K in the roots, trunk and cordons at budbreak in Chenin blanc vines at a 2.44 × 2.44 spacing

Table 6.4. *Nitrogen and potassium budgets for Thompson Seedless grapevines grown in the San Joaquin Valley of California*

	Vine part	Amount[a] (g vine^{-1})
Nitrogen		
Requirements	Leaves	35
	Stem[b]	10
	Clusters	30
	Total	**75**
Losses	Shoot trimming[c]	8
	Fallen leaves	20
	Prunings	15
	Fruit harvest	30
	Total	**73**
Potassium		
Requirements	Leaves	12
	Stems	26
	Clusters	45
	Total	**83**
Losses	Shoot trimming	5
	Fallen leaves	8
	Prunings	11
	Harvest	45
	Total	**69**

[a] Vine density: 1120 vines ha^{-1}.
[b] The stem is the main axis of the shoot.
[c] The shoots were trimmed near to the ground prior to harvest to facilitate soil preparation for drying the grapes for raisin production.
Adapted from Williams (1987) and Williams *et al.* (1987).

(1680 vines ha^{-1}) was equivalent to 215 and 124 kg ha^{-1}, respectively (see Table 4.3). The mineral nutrients in the permanent structures of the vine may be remobilized to support new season's growth, including growth of new roots, when the uptake from the soil is insufficient to meet the current demand.

The seasonal demand for mineral nutrients is greatest during the period of maximum shoot growth; it is also substantial during the period of rapid cluster growth (see Figs. 4.8 and 4.9) (Alexander, 1958; Conradie, 1980, 1981a; Lohnertz, 1988; Williams, 1987; Williams *et al.*, 1987). The uptake and storage of nutrients in the permanent structures of the vine can take place throughout the growing season. The period after harvest is also important from a nutritional viewpoint because the nutrient re-

Table 6.5. *The amounts of mineral nutrients removed when one tonne of fruit is harvested*

Mineral nutrient	Amount (kg t^{-1})		
	average[a]	high	low
N	1.46	2.06	0.90
P	0.28	0.39	0.22
K	2.47	3.69	1.59
Ca	0.50	0.93	0.17
Mg	0.10	0.16	0.05

[a] Averages for N and K were from seven of the studies. Averages for P, Ca and Mg were from three of the studies. The 'high' and 'low' columns represent the highest and lowest values from published reports.
Data are from Conradie, 1980, 1981a; Lafon *et al.*, 1965; Lohnertz, 1988; Marocke *et al.*, 1976; Williams, 1987; Williams and Biscay, 1991; Williams *et al.*, 1987; see also Table 4.3.

serves in the roots and trunk that have been depleted during the current season are replenished.

FERTILIZATION OF VINEYARDS

The need to apply fertilizers to vineyards is indicated by the occurrence of deficiency symptoms, low vine vigor, reductions in yield, or the results of vine tissue analysis. Once the decision to fertilize has been made, the amount to apply is affected by soil characteristics and by the severity of the deficiency. The soil texture and the amount of organic matter determines the capacity of the soils to bind minerals to various particles in the profile. These particles can have either positive or negative charges, and they bind mineral ions of the opposite charge. Sandy soils, which have little capacity to bind mineral nutrients, are more prone to leaching of minerals (the movement of the minerals in the soil profile) below the root zone (deep percolation) than heavy soils. Therefore, the amount of fertilizer applied to a vineyard with sandy soil is usually greater than that applied to vineyards with loam or clay soil because a large proportion of the fertilizer may be lost to deep percolation.

Soil pH also determines the extent of mineral ions available to the vine. If the soil becomes too acid (pH < 6.5), excess hydrogen ions will replace cations such as Ca, K, Mg and Na, and these cations may be leached downward through the profile. In addition, the solubilities of aluminum and manganese, which are toxic to plants, are increased. The continued use of large amounts of N fertilizers may also increase the acidity of the soil (Conradie and Saayman, 1989). The application of lime (calcium carbonate) to soils is used to increase the pH of an acid soil. If the soil becomes too alkaline (pH > 8.0) some mineral nutrients may become unavailable to the vines, and there may be an increase in the availability of toxic borate ions. The ideal soil pH for growth of most cultivated plants is within the range from 6.0 to 7.0; however, a grapevine will grow adequately in soils with a pH as high as 8.0 unless it is sensitive to high lime.

The cultural practices employed in a vineyard also affect the amount and kind of fertilizers to be applied. To a large extent the type of irrigation system that is used controls the distribution of roots within the soil profile. Vines that are irrigated with a drip or low-volume irrigation system have root systems that are concentrated under the emitters. Thus, the fertilization of vines with drip irrigation (fertigation) is highly efficient because the fertilizer is placed directly beneath the emitter in the zone where most of the roots are to be found. This may result in a reduced requirement for applied fertilizers. The use of a leguminous cover crop between rows may lessen the need for N fertilizers. However, the need for other fertilizers may increase because the cover crop may compete with vines for mineral nutrients.

In the past, recommendations for N fertilization in the vineyard were for applications during the dormant period before budbreak. This was to ensure that adequate N would be available in the profile for vine growth early in the season. However, a study using ^{15}N-depleted ammonium sulfate showed that dormant season application is inefficient because much of the N is leached from the soil by winter rains before vine growth commences (Peacock et al., 1982). Once vine growth has begun in the spring much of the vine's N supply comes from the remobilization of reserve N in the roots, trunk and cordons (Conradie, 1986). Accordingly, the dormant part of the growing season is unsuitable for application of N fertilizers. When vines growing in sand culture were fertilized with N, either in the spring (at anthesis) or in summer (before véraison), approximately 43% of the fertilizer N was translocated to the clusters and 22% was found in the permanent structures at the start of the next growing season (Conradie, 1986). The fertilization of vines during the postharvest period results in the highest concentration of fertilizer N in the vine at budbreak the next season (Conradie, 1986; Peacock et al., 1989). Postharvest fertilization with N is beneficial only if applications are made while the leaves remain on the vine.

Most soils are comparatively high in total potassium, but only a very small percentage of this macroelement is in a form that is available to plants (such as K ions in solution or as exchangeable K adsorbed onto soil colloids). The fertilization of vineyards with K can be effective with applications at any time during the growing season. This can be accomplished by applying K either as a single application or by use of numerous smaller amounts applied through the drip system. Over-fertilization with K may result in high pH and other problems concerned with the quality of grape juice and wine (Hale, 1977; Morris *et al.*, 1980).

Phosphorus deficiency rarely occurs in vineyards throughout the world, but Cook *et al.* (1983) found that P deficiency was evident in the foothills of the Sierra Nevada and hillsides of the California coastal mountain range. This deficiency of P was only found in areas with acid soils.

A deficiency of Zn or Fe can be corrected with both soil- and foliar-applied fertilizers in vineyards. Under conditions of Zn deficiency, an application of a foliar Zn fertilizer shortly before anthesis will increase the number of flowers that set (Christensen, 1980).

Vineyard water use (evapotranspiration)

Water use in a field situation involves water loss through transpiration from the crop and evaporation from the soil. Collectively, these two components of water loss are termed evapotranspiration (ET). The ET of a particular crop is dependent upon the stage of plant growth, the evaporative demand of the atmosphere throughout the season and soil water availability. A recent review of the literature on irrigation principles, soil and water relations and ET has been made by Stewart and Nielsen (1990).

The production of grapes in many areas of the world is limited by water availability, and lack of water reduces yield and fruit quality. In certain areas of Europe, the use of supplemental water is forbidden by law for the production of premium wine grapes. It is widely believed that irrigation reduces the quality of the wine. Spain leads the world in acreage devoted to grape production, but it ranks fourth in wine production owing to low yields. Part of the reason for these low yields is the lack of irrigation.

The use of water by a vineyard is characterized by low demand early in the growing season and after harvest and by high demand when canopies are fully developed. Evaporative demand is also less early and late in the year than at the height of the growing season.

Cultural practices, especially the type of trellis system, also affect the amount of water used by a vineyard. Large and high trellis systems generally produce greater yields than trellis systems which produce a smaller total leaf area, and it is often assumed the larger systems use more water. However, Van Zyl and Van Huyssteen (1980) found that vineyard ET of bush-trained vines is greater than that of vines in which the canopy is spread out by the trellis system. The greater ET of the bush vines was due to higher ambient temperatures, to more air movement among the vines and to less shading of the soil. Girdling grapevines decreases the water use of vines for approximately one month after the girdling takes place (Williams and Matthews, 1990). Girdling causes a decrease in stomatal conductance and this is the reason for the decrease in water use. Irrigation frequency also has an effect on vine water use. If the soil water is depleted to the point that the vines are stressed, the use of water by the vines will decrease (Grimes and Williams, 1990). An increase in crop load will also increase vineyard ET (Van Zyl, 1987). Lastly, any cultural practice or insect damage that diminishes leaf area will also decrease vine water use.

Vineyard water use (ET_c) for mature vines in the semiarid southwestern USA and in Australia, where climatic conditions are similar to those in California, varies from 650 to 800 mm (Williams and Matthews, 1990). Daily vineyard ET in these climates can be as great as 10 mm (Grimes and Williams, 1990). In the desert regions of California where table grapes are grown, vineyard ET was estimated to be 1110 mm per season (Jacob, 1950). Van Zyl (1988) calculated that ET_c of furrow-irrigated vines in the Oudtshoorn of the Little Karoo region of South Africa was 581 mm.

ET_c of vineyards during vine establishment is less than that of a mature vineyard. With the use of a weighing lysimeter, measured water use for Thompson Seedless vines in the San Joaquin Valley of California during the first three years of vineyard establishment was 300, 400 and 590 mm, respectively. The reference crop ET was approximately 1250 mm year^{-1} during the three years (unpublished data of L.E. Williams, D.W. Grimes, L.P. Christensen and C.J. Phene). Vineyard ET during the third year, when only two 12-bud fruiting canes were retained at pruning, was almost the same as that expected of a mature vineyard where more fruiting canes are retained at pruning.

Vineyard irrigation practices

EFFECTS ON DRY MASS PARTITIONING WITHIN THE VINE

The total amount of dry mass partitioned among the permanent structures of the vine is decreased when the amount of water available to the vine is limited owing to deficit irrigation. The amount of dry mass found

Table 6.6. *The effect of two irrigation treatments on the partitioning of dry mass and non-structural carbohydrates (NSC) among the permanent structures of Chenin blanc grapevines grown near Lodi, California*[a]

Irrigation treatment	Vine organ		
	cordons	trunk	roots
Dry mass ($g\ vine^{-1}$)			
100% ET_c	3668[b]	2383	3382
52% ET_c	2717	1985	2339
% reduction	26	17	31
NSC ($g\ vine^{-1}$)			
100% ET_c	634	452	827
52% ET_c	416	315	562
% reduction	34	30	32

[a] The vines were irrigated at 100 and 52% of calculated vineyard ET (ET_c) for a four year period. The vines were harvested prior to budbreak before the beginning of the fifth year. (The assistance of T. Prichard and P. Verdegaal is gratefully acknowledged.)
[b] Each value is the mean of four replicates.

in the cordons and roots was reduced by approximately 30% after vines had been irrigated at 52% of vineyard ET as compared with those irrigated at 100% ET (Table 6.6). The reduction in dry mass of the cordons was less than 20% when the two treatments were compared. In addition to the decrease in dry mass, the amounts of non-structural carbohydrates decreased by approximately 30% at the lower irrigation level. Deficit irrigation also resulted in an increase in the amount of dry mass per unit leaf area of the vines receiving less water; and this result was due to an increase in structural components (cell walls) of the leaf and not to an increase in non-structural carbohydrates (Williams and Grimes, 1987). The most marked change in vine growth with supplementary irrigation is an increase in the growth of the shoots and an increased leaf area per vine (Williams and Matthews, 1990). The increase in shoot growth with irrigation also results in an increase in pruning masses (Smart and Coombe, 1983; Williams and Matthews, 1990).

The reproductive growth (final yield) of vines increases with water applications (Smart et al., 1974; Van Zyl, 1984; Grimes and Williams, 1990). As vineyard water use increased, so did vineyard yield for Thompson Seedless grapevines grown in the San Joaquin Valley of California (Fig. 6.5). This increase in yield was linear from approximately 40% ET_c up to full ET_c.

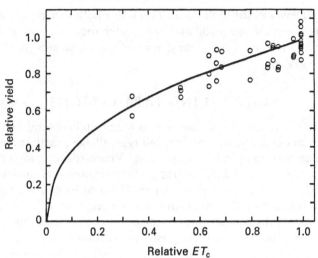

Fig. 6.5. A water production function ($Y_r = 0.976\ ET_c^{0.409}$; $R^2 = 0.72$) of Thompson Seedless grapevines relating relative yield Y_r (observed yield/maximum yield) to relative crop evapotranspiration (observed ET_c/maximum ET_c). From Grimes and Williams (1990). Reproduced with permission

ROOT DISTRIBUTION AND SOIL WATER DEPLETION

Regardless of irrigation regime, little root growth occurs before bud-break or during the middle part of the growing season when demand for water by the vine is at a maximum. Fewer new roots were found in the driest soil when the vines were allowed to deplete soil water to 25% of the available soil moisture content (Van Zyl, 1984).

The type of irrigation system also affects the distribution of roots in the soil profile. With drip irrigation approximately 50% of the total number of roots in the profile are in the 0–20 cm soil horizon profile. With furrow irrigation, less than one percent of the total roots are found in the top 20 cm (Araujo, 1988). Root distribution patterns are similar under systems of furrow irrigation and 'border irrigation', where the inter-row areas are flooded (Van Zyl, 1988). The water depletion patterns with these two systems of irrigation were similar in that 40, 30, 20 and 10% of the water extracted occurred in the upper 25%, the second, third and fourth quarters of the root zone, respectively. Vine spacing within the vineyard also affects the water extraction pattern. The closest spaced vines deplete soil water more rapidly than the widest spaced vines at all levels in the soil profile. This is due both to a greater density of roots per unit volume of soil and to a greater angle of penetration (i.e. root growth more vertical) as the density of vines increases (Archer and Strauss, 1989).

SCHEDULING IRRIGATIONS

The timing of irrigation of vineyards is based upon several factors including water cost and availability, soil type, tillage practices and local regulations governing irrigation practice. Vineyards with high density plantings (Archer and Strauss, 1989), large canopies (Peacock et al., 1987) and which use cover crops require frequent irrigations. Effective rooting depth in the soil, evaporative demand and type of grape product (wine grapes, table grapes, raisin grapes) also determine the amount and frequency of irrigation. At present, irrigation by most growers is done on a calendar basis or in response to visual assessments in the vineyard. In most irrigation districts (agricultural areas receiving irrigation water from a specific source), growers must wait their turn for irrigation water to become available, and rosters and administrative factors tend to determine *when* grapevines are irrigated. Ideally, irrigation frequency should be based on quantitative measures of either vine or soil water status and its relation to productivity or quality of the fruit.

Recent developments in remote sensing and its application to the assessment of plant water requirements (Idso *et al.*, 1977; Jackson *et al.*, 1981) provide a new technology for precision irrigation of crops. This technology involves determining the difference in ambient and canopy temperature (which is measured with an infrared thermometer) and its dependence upon atmospheric vapor pressure deficit (Idso *et al.*, 1981). The values generated with this method are termed 'crop water stress index' or CWSI (Jackson *et al.*, 1981). This technology is being developed for use in vineyards. Van Zyl (1986) determined that canopy temperature is linearly related to soil water content, and Grimes and Williams (1990) found that the CWSI measured in a Thompson Seedless vineyard in a semiarid environment is linearly related to vineyard yield (Fig. 6.6). To be effective, vineyard irrigations should be scheduled before the CWSI reaches 0.3. The advantage of infrared thermometry is that the measurements can be made routinely and nondestructively. A disadvantage is that IR measurements cannot be taken when it is windy or cloudy. Further, the absolute values of CWSI and their meaning with regard to vine water status can differ from cultivar to cultivar.

Scheduling of irrigations is also based on (i) the use of meteorological

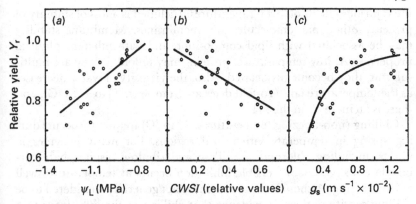

Fig. 6.6 The relations between relative yield (Y_r) and various measures of plant water status of Thompson Seedless grapevines. (a) Leaf water potential (Ψ_L): $Y_r = 1.29 + 0.409\ \Psi_L$, $R^2 = 0.58$. (b) Crop water stress index: $Y_r = 0.982 - 0.312\ CWSI$, $R^2 = 0.43$. (c) Stomatal conductance (g_s); $Y_r = 1.015\ \exp(-0.088\ g_s^{-1})$, $R^2 = 0.61$. From Grimes and Williams (1990). Reproduced with permission.

data to estimate the vine water use and (ii) knowledge of the amount of water available in the soil profile between irrigations. For this method to be useful it is necessary to determine the water holding capacity of the soil within the vineyard. In addition, crop coefficients need to be developed for the vines throughout the growing season for the specific cultivar, trellis type and location. The crop coefficient (k_c) is the fraction of the amount of water used by a vineyard relative to water use by a reference crop such as grass. Reference crop ET is designated: ET_0. Thus k_c equals ET_c divided by ET_0. Owing to the development of the vine's canopy throughout the season, the k_c of vines will increase from spring to harvest and then decrease sometime after harvest.

The type of irrigation delivery system may determine, in part, the amount and frequency of the irrigation regime. Drip irrigation is more efficient than other irrigation delivery systems (Peacock et al., 1977b; Smart et al., 1974). The efficiency of drip irrigation can be increased by use of frequent applications, either daily or every other day (Goldberg et al., 1971; Smart et al., 1974). This increased efficiency is due to the conservation of water that would be lost to deeper soil layers when greater amounts of water are applied at less frequent intervals.

Winterkill and frost damage

Grapevines in temperate climates experience low temperatures and freezing stress from late fall to early spring. Several factors contribute to the injurious effects of low temperatures. Included is a loss of stability of proteins, other macromolecules and membranes. Membrane stability may be associated with lipid composition and with inherent physical properties. At low temperature, proteins may unfold; if they are multi-subunit, they become dissociated. Also, the viscosity of water decreases as the temperature falls (10-fold decrease from $+25\,°C$ to $-25\,°C$), and there is an increase in the pH.

Chilling (non-freezing temperatures, $1-12\,°C$) frequently occurs during spring in temperate viticultural regions. For many enzymes in plants, including chloroplast ATPase and ribulose bisphosphate carboxylase–oxygenase, reversible inhibition occurs at temperatures well above freezing. Although the grapevine is not generally considered to be chilling-sensitive, there is evidence that chilling in the light leads to a temporary inhibition of photosynthesis (Balo *et al.*, 1986).

Freezing stress, whether winterkill or frost damage, occurs when tissue water undergoes freezing. Freezing temperatures damage plant tissues by physically rupturing cells, disrupting membrane function, and irreversibly denaturing enzymes by dehydration. This damage occurs when the water present in plant tissues freezes and expands in volume. The apoplastic solution is more dilute than that of the symplast and it freezes first. Generally, freezing in the apoplast is not injurious, but it takes water out of solution and thereby lowers the apoplastic water potential. In turn, this causes water to move from inside the cell to the apoplast. Damage occurs when sufficient water has left the cells to cause dehydration.

There are significant developmental changes in the sensitivity of buds to spring frost; dormancy is an important acclimatizing process that diminishes susceptibility to frost and freezing injury. In Germany, Müller-Thurgau, Sylvaner, Gewürztraminer, and Muscats are more susceptible to frost damage than White Riesling and Elbing. In the United States, Chardonnay often shows more injury to spring frosts than other cultivars. Frost damage is sometimes called first degree (vegetative shoot tip only) and second degree (flower cluster also damaged or killed).

Freezing injury, or winterkill, occurs when permanent, overwintering vine parts are damaged by low temperatures. Temperatures below $-15\,°C$ are often required to cause significant freezing injury because acclimation to freezing temperatures is a normal aspect of vine physiology. As with frost, susceptibility to winterkill varies considerably during the season. The acclimation process begins after shoot growth ceases, i.e. well in advance of freezing temperatures. Acclimation is induced by short days in most woody species, but photoperiodic induction is less clear in *Vitis*. Brief exposures to o °C accelerate acclimation, and maximum hardiness is obtained after exposure to − 5 °C.

The freezing of different pools of water can be observed in exotherms (release of latent heat in the phase transition of water from liquid to solid upon freezing) as tissues are subjected to decreasing temperatures (Pierquet *et al.*, 1977). As acclimation proceeds, susceptibility to freezing temperatures, and corresponding changes in these exotherms, can be observed. Acclimation to temperatures as low as − 40 °C occurs in *V. riparia*. This observation, and the observation of low temperature exotherms at − 35 to 40 °C, have led to the suggestion that an important aspect of acclimation may involve the supercooling of tissue water to below freezing temperatures, i.e., a freeze avoidance. The theoretical limit of deep supercooling is approximately − 45 °C, and this is consistent with the low temperatures regularly observed at the northern limit of occurrence of *V. riparia*. Because other exotherms occur at higher temperatures, the freezing tolerance may be largely due to the capacity to compartmentalize water that is susceptible to freezing.

Also during acclimation, tissue water content decreases, soluble proteins in the bark increase, the thermal stability of several enzymes increases, and membrane permeability increases. In general, poor growing conditions during the season inhibit the acclimation process. Acclimation is promoted by, or correlated with, shoot cane exposure to sunlight, periderm development, and low relative water content. Heavy fruit loads or defoliation inhibit acclimation, probably through losses in available photosynthate. Cold acclimation is rapidly lost upon exposure to high temperatures and the loss of acclimation is closely correlated with the recovery of tissue water content in the spring. Withholding irrigation water to promote moderate water deficits has been proposed as beneficial for cold acclimation, and there is evidence supporting this hypothesis in other species.

Species vary greatly in their susceptibility to winterkill. In order of decreasing susceptibility, vines are classified as follows: *V. rotundifolia* > *V. vinifera* > 'French hybrids' > *V. labrusca* and *V. riparia*. Tissues also vary in tolerance to freezing temperatures, in order of decreasing susceptibility: primary latent buds > secondary latent buds > cane xylem. The primary latent bud can be as much as 10 °C less hardy than the secondary bud (Stergios and Howell, 1977).

Use of plant growth regulators in the vineyard

The use of plant growth regulators in vineyards to control or alter growth has been studied for many years (Weaver, 1972). Growth regu-

Table 6.7. *The effect of trunk girdling and GA₃ application at fruit set on berry mass (g berry⁻¹) of Thompson Seedless grapevines*

Girdle	GA₃ applicationᵃ				Mean effect of girdle
	none	vine	shoots	clusters	
No	2.00	2.70	2.38	3.16	2.56
Yes	2.83	3.55	3.27	3.85	3.35
Mean effect of GA₃	2.42	3.12	2.77	3.50	—

ᵃ Gibberellic acid was applied to the entire vine (vine), to shoots only (shoots), or to clusters only (clusters).
Harrell and Williams (1987).

lators are used extensively in the production of table grapes. Gibberellic acid (GA_3) has been used since the 1960s to increase the size of seedless grapes. The standard practice is to spray GA_3 at fruit set, approximately two weeks after anthesis, in order to increase berry size. GA_3 increases berry size by inducing an increase in cell numbers and elongation of the cells (Sachs and Weaver, 1968). The application of GA_3 also alters the accumulation of photosynthates; the organs that are sprayed tend to accumulate the most sugars (Weaver *et al.*, 1969). The optimal response to GA_3 is dependent upon the time of application, the concentration, and the portion of the vine that is sprayed (Table 6.7).

A GA_3 application pre-anthesis can lengthen the rachis in some cultivars. An application of this phytohormone at anthesis thins the cluster by inducing floral abscission and/or increasing 'shot berries'. This results in clusters that are not excessively compact and in which the potential for fungal rots is reduced. Vines are also girdled at fruit set to further increase the size of seedless grapes. The application of GA_3 and girdling at fruit set have additive effects on berry size (Table 6.7). A potential drawback to an application of GA_3 is that it causes bud necrosis in certain table grape cultivars (Ziv *et al.*, 1981); this results in fewer clusters per vine in the next growing season (Harrell and Williams, 1987).

Other uses of GA_3 in the vineyard include applications to Black Corinth (syn. Zante currant) vines during anthesis to increase the size of the fruit for the production of currants. Gibberellin has been used on some normally tight-clustered wine grape varieties to produce lax bunches, which are less susceptible to *Botrytis*. The application of GA_3 at anthesis causes significant flower abscission in these cultivars; however, subsequent bud fruitfulness is severely reduced.

Another plant growth regulator used extensively in grape production is ethephon (2-chloroethylphosphonic acid), an ethylene-releasing agent. The ethylene so produced affects enzyme activities, mineral status and other processes within grapevines (Szyjewicz *et al.*, 1984). Ethephon has been used to promote color development and sugar accumulation in red and black-colored berries of numerous cultivars. Ethephon is normally applied to vines at véraison, and it may replace the girdling of table grape cultivars at véraison to enhance coloring (Peacock *et al.*, 1977a). The numerous grape cultivars in which fruit color and sugar accumulation respond positively to applications of ethephon are listed by Szyjewicz *et al.* (1984).

The use of ethephon to induce abscission of flowers, berries or leaves has been used with varying degrees of success. When used to reduce the number of berries there has been uneven thinning (Weaver and Pool, 1971) or complete crop loss (Szyjewicz and Kliewer, 1983). The ease of abscission of berries from the rachis is affected by the time of application and environmental conditions. The use of ethephon to aid in the harvest of Muscadine grapes (*Muscadinia rotundifolia*), which do not form an abscission layer at maturity, and in standard *V. vinifera* cultivars has been studied by Phatak *et al.* (1980) and El-Zeftawi (1982) with variable results. In grape production areas that do not experience low winter temperatures, the application of ethephon at high concentrations (4000–5000 ppm) causes defoliation and enhanced budbreak (Corzo, 1982).

Excessive vegetative growth of grapevines is associated with dense canopies, low bud fruitfulness and inferior fruit quality. This type of growth also increases the possibility of infection by various pathogens. Ethephon has been shown to inhibit both the growth of the primary shoot and lateral bud outgrowth for a period of about two months after application (Lavee, 1987). The inhibition of lateral bud growth is important because summer pruning to control vegetative growth releases lateral buds from apical dominance and regrowth from lateral buds results in even more vegetative growth.

Paclobutrazol (2RS, SRS)-1-(4-chlorophenyl)-4, 4-dimethyl-2 (1, 2, 4-triazol-l-yl)-pentan-3-ol, a new plant growth regulator, is an effective vegetative growth retardant for tree fruit species. It has been used in trials on pot-grown and field-grown *V. vinifera* and *V. labrusca* cultivars, both as a foliar spray and by soil application (Intrieri *et al.*, 1986; Ahmedullah *et al.*, 1986; Williams *et al.*, 1989). Both methods of application reduced vegetative growth of the vines. However, the adverse effect of paclobutrazol on the yield of Thompson Seedless grapevines, owing to reduced numbers of clusters, indicates that it would be of little practical use on this cultivar when applied through the soil (Williams *et al.*, 1989).

Many grape-producing areas in the world lack adequate chilling during the winter months to ensure regular budbreak. Shoot emergence in these areas is irregular and many of the buds fail to grow. Under these conditions, budbreak of grapevines can be manipulated by the use of several chemicals or growth regulators (Kuroi *et al.*, 1963; Shulman *et al.*, 1983; Weaver *et al.*, 1974). The most successful of these compounds is hydrogen cyanamide (H_2CN). When cyanamide is applied to vines shortly after pruning, it induces earlier and more uniform budbreak. It is widely used in the tropics and desert regions where chilling is inadequate for uniform budbreak.

Studies in Israel have demonstrated that a class of synthetic growth inhibitors called morphactins will chemically 'girdle' a grapevine (Shulman *et al.*, 1986) when painted on the bark. Berry size and fruit maturation were increased with bark applications of this growth regulator at berry set and at six weeks before harvest. However, this chemical girdle was less effective than a conventional mechanical girdle.

The effectiveness of the growth regulators, pesticides and foliar nutrient sprays described above depends upon the material penetrating the cuticle of the vine organ or formation of a continuous film over the surface. Substances called adjuvants are often added to the spray solution to ensure better penetration or to act as a spreader or 'sticker'. An adjuvant is defined as any substance added to the spray tank, separate from the specific chemical being used, that will improve the performance of that chemical. Adjuvants also may be called surfactants or wetting agents. The uptake of spray solutions by the grapevine may occur through the stomata or through other channels in the cuticle. Many of the adjuvants used in spray applications may break down the waxy cuticle of the leaf or berry, thereby facilitating the entry of the chemical. The use of such surfactants may also facilitate the entry of disease organisms into the grape berry (Marois *et al.*, 1987).

Biotic stress: pests

Cultivated grapes are highly susceptible to infestation by pests and to attack by pathogenic microorganisms. This section gives a brief introduction to the principal pests and diseases of *vinifera* grapes and to modern strategies in crop protection. Information presented here is not meant to be comprehensive; those requiring further details should consult specialized texts in entomology, plant pathology and plant protection.

ABOVE-GROUND PESTS

Viticulture is a monoculture system based on few genotypes; these circumstances are highly favorable for pest infestations. The above-ground parts of grapevines are subject to attack by a very wide range of insects and mites. *Lepidoptera* (caterpillars), *Coleoptera* (beetles), *Orthoptera* (grasshoppers), *Homoptera* (scales, aphids, leafhoppers), *Hemiptera* (bugs), *Diptera* (flies), *Dermaptera* (earwigs), *Isoptera* (termites) and *Acarina* (mites) are all represented. The relative importance of the various pest species affecting leaves, stems, flowers and fruits differs very greatly among grape growing countries and among regions within countries. On a global scale, a list of the most important pests of the aerial organs of grapevines would be very long, and it would require frequent revision because new pests arise and old pests decline in importance in response to various biotic, climatic and managerial factors.

In contrast, the major pests of grapevine roots belong to a few taxa, and the same pests are of great importance in all countries where viticulture is of commercial significance. These pests are (i) grapevine phylloxera, a relative of the aphids, and (ii) plant parasitic nematodes.

SOIL PESTS

GRAPEVINE PHYLLOXERA

Phylloxera, a root-louse, which feeds on the roots and leaves of grapevines, is the scourge of viticulture. It has been a major pest in all grape growing countries for at least a century. The former scientific names of the organism concerned were *Phylloxera vitifoliae* FITCH and *Phylloxera vasatrix* PLANCHON. It has now been assigned to a different genus and re-named *Daktulosphaira vitifoliae* FITCH (*Homoptera*: *Phylloxeridae*), but 'phylloxera' has been retained as a common name. Phylloxera is native to the United States of America and it is widely distributed to the east of the Rocky Mountains. It is an introduced pest in California.

The life cycle of phylloxera is complex (Fig. 6.7). In brief, there are two main phases: (i) a root feeding phase known as *radicicole* and (ii) a leaf-feeding phase known as *gallicole*. The winged form of phylloxera, which gives rise to the sexual forms, occurs in Europe but it is not found in California. Phylloxera in California are oviparous, parthenogenetic females.

Phylloxera have sucking mouth-parts, and they produce characteristic lesions on the roots of grapevines known as nodosities and tuberosities.

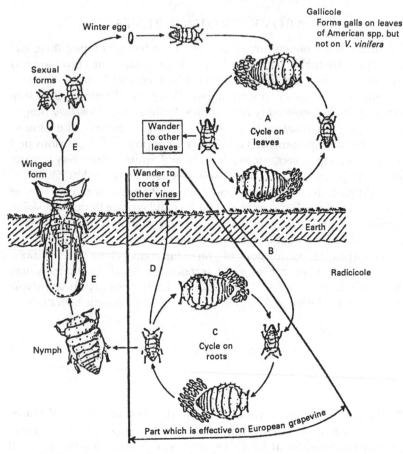

Fig. 6.7. Diagram of the life cycle of phylloxera (*Daktulosphaira viti-foliae*). From Winkler *et al.* (1974). Reproduced with permission. (Copyright © 1974 The Regents of the University of California.)

Nodosities are small swellings due to subepidermal callus formation. The feeding of phylloxera induces cell division, but it is not yet clear if the formation of nodosities results from the secretion of phytohormones by the insect or whether phylloxera triggers the plant to overproduce growth substances which, in turn, leads to cell division and lesion formation. Tuberosities are large nodosities and are characteristic of susceptible host plants. Nodosities can occur on the roots of genotypes that are classified as phylloxera-resistant for viticultural purposes, and a distinction needs to be made between tolerance and true resistance of grapevine root systems to phylloxera.

The roots of susceptible, phylloxera-infested grapevines become distorted by nodosities and tuberosities, the regeneration of new roots is inhibited, and the efficiency of water and mineral uptake by the root system is greatly impaired. In due course the affected vines become weakened and die. Of great viticultural significance is the fact that the roots and leaves of *Vitis vinifera* and many North American *Vitis* have different resistances and susceptibilities to phylloxera:

	Vitis vinifera	American *Vitis*
LEAVES	RESISTANT	SUSCEPTIBLE
ROOTS	SUSCEPTIBLE	RESISTANT

The gallicole forms do not attack the leaves of *Vitis vinifera* but they form characteristic galls on the leaves of a wide range of American species. These galls do not seem to cause significant damage to the infested vines. Most species of *Vitis* native to North America are tolerant or resistant to the radicicole form of phylloxera. A few nodosities can be observed in many American species but tuberosities do not occur. The roots of *Vitis vinifera* are highly susceptible to the radicicole form and produce both nodosities and tuberosities. There is rapid growth of phylloxera populations on *vinifera* roots. The pest is highly damaging and infestations lead to the death of the vine.

The accidental introduction of phylloxera into Europe in the 1860s was a major disaster. Vines from the east coast of the United States were sent to France as museum specimens. The vines concerned, which were either North American species or interspecific hybrids, were infested with phylloxera. The insect soon established itself in the vineyards of France, where the vines were exclusively the highly susceptible *Vitis vinifera*.

The spread of phylloxera was very rapid. By 1900, 75% of French vineyards were affected and this led to great rural poverty and distress. Subsequently, phylloxera spread to Italy, Spain, Portugal and Yugoslavia and to all other grape growing regions of Europe. Phylloxera reached California in 1873. Later, it appeared in South America, South Africa, Australia and New Zealand.

The solution to the phylloxera problem was to use American species or interspecific hybrids as rootstocks for *Vitis vinifera*, i.e. to graft susceptible scions onto resistant root systems. The breeding of phylloxera-resistant rootstocks is described in detail in Chapter 7. Phylloxera-resistant rootstocks are the foundation of European and American viticulture. There are few places in the grape-growing world that are phylloxera-free and where cultivars of *Vitis vinifera* can be grown on their own roots. Exceptions are the main grape-growing districts of

Australia where phylloxera has been excluded, so far, by a combination of quarantine regulations and geographical isolation from regions which became phylloxera-infested in the nineteenth century. In addition, there is still considerable acreage in California in which vines are planted on their own roots.

Plant parasitic nematodes are highly damaging pests of grapevine roots. Species of the genera *Meloidogyne* (root knot nematodes) *Pratylenchus* (root lesion nematodes), *Tylenchulus* (citrus nematode), and some others of less significance, reduce grapevine vigor and production by the damage they inflict during feeding. Root knot nematodes induce the formation characteristic of galls, which act as competing sinks for photosynthate. With all species of parasitic nematodes, injury to the root system through feeding results in reduced production of new roots and to reduced uptake of water and nutrients. Root knot nematodes, in particular, thrive in sandy soils and are known to move through the profile and into drainage water. This characteristic makes root knot nematodes a serious pest of irrigated viticulture.

Dagger nematodes of the genus *Xiphinema* cause damage to grapevine roots but are more important as vectors of virus disease. *X. index* is a vector of grape fanleaf virus (GFLV), the causal organism of fanleaf degeneration. *X. index* is also a vector of yellow mosaic and of several other viruses. Other species of *Xiphinema* are thought to be vectors of Arabis mosaic and Hungarian chrome mosaic.

CONTROL MEASURES: INTEGRATED PEST MANAGEMENT

During the past 30 years there have been major developments in the chemical control of insects and mites, and viticulturists have access to a wide range of highly effective spray materials. However, it has become clear that there are serious risks associated with over-reliance on chemical methods of plant protection. Secondary pest outbreaks and the development of resistance are common problems, but the main concerns are the occurrence of toxic residues in foodstuffs, and contamination of the environment. In most countries, these concerns have prompted stringent legislation on food safety standards and environmental protection. In turn, this has led to (i) a slowing down in the rate of release of new insecticides and acaricides because of the very high costs of development and testing; and (ii) much interest in alternative strategies for pest control. In viticulture and other forms of fruit production, integrated pest management (IPM) is a new approach to crop protection.

IPM involves the use of many management techniques (including highly judicious use of chemical protectants) to keep pest populations below a level which causes economic loss. The goal of IPM is to prevent economic injury to the vine rather than to eradicate the pest. In many instances, the damage caused by pests is not as serious as it appears. Vigorous grapevines can lose a large area of leaves without incurring economic damage. In California, for example, wine and raisin grapes grown with irrigation can tolerate a 20% loss of leaf area from the time of fruit set until July without detrimental effects on production and fruit quality. In the month preceding harvest, leaf area can be reduced by 50% without affecting yield, and there is no reduction in the next year's crop if vines are defoliated in September (production may be affected if early defoliation occurs over several years). Under these circumstances, the use of repeated applications of toxic insecticides to achieve pest-free vines serves no useful purpose; it is counterproductive because such severe, untargeted treatments eliminate useful predators as well as pests.

Breeding crop plants for pest resistance, and the use of resistant genotypes, are important elements of IPM systems. This is more easily achieved in herbaceous annual crops than in woody perennials. As explained in other chapters, plant breeding at the level of the scion is subject to many difficulties and has had relatively little impact on viticulture (an exception would be the introduction of new table grape cultivars). In contrast, the breeding or selection of pest-resistant rootstocks for grapes is among the earliest and best examples of biological control. The importance of grapevine rootstocks that are resistant to phylloxera and nematodes is emphasized by new constraints on the use of soil-applied insecticides and nematicides. In some grape growing countries, persistent molecules which move through the soil profile, and which are potential contaminants of ground water, are being phased out or are already prohibited by law.

All insects and mites have natural enemies or predators, which act as a brake on population growth. The importance of these beneficial organisms was not fully appreciated until the advent of synthetic pesticides and non-selective control measures. The use of highly toxic, broad-spectrum, plant protective chemicals led to outbreaks of secondary pests because of the destruction of predators, and to resurgence in the populations of target species owing to the development of resistance. In many crops there have been spectacular successes with classical biological control, the deliberate introduction of predators of the target species, and this approach is being applied to viticulture. In California, major research programs are in progress to introduce and establish natural enemies of two important pests of grapes, variegated leafhopper (*Erythroneura variabilis*, BEAMER (*Homoptera*)) and western grapeleaf skeletonizer (*Harrisina brillians* BARNES and McDUNNOUGH (*Lepidoptera*)). The predators concerned are *Anagrus epos* (GIRAULT), a wasp that parasitizes the eggs of the variegated leafhopper, and a wasp (*Apanteles harrisinae* MUESBECK), and a fly (*Ametadoria (Sturmia) harrisinae* (COQUILLET)) both of which parasitize the western grapeleaf skeletonizer.

Successful IPM is founded on a systems approach to the control of insect or mite infestations. This is well illustrated by strategies for control of leafhoppers in irrigated vineyards of the San Joaquin Valley of California. A coarse control of leafhopper infestations can be achieved by managing grapevine vigor through irrigation scheduling and supply of nutrients. Excessive vigor favors the growth of leafhopper populations; it can be avoided by withholding water and nitrogen so that there is no active extension growth after midsummer. The *Anagrus* parasite does not overwinter on grapevines but establishes itself on blackberries (*Rubus* sp.), particularly in riparian habitats. The French prune (*Prunus domestica* L.) is an alternative overwintering host. If no natural overwintering sites for *Anagrus* are available, grape growers are encouraged to plant prune trees in, or near to, their vineyards so as to facilitate the rapid buildup of *Anagrus* populations early in the growing season.

Cover crops and volunteer grasses, which are mowed periodically during the growing season, can act as insectaries for general predators of leafhoppers (lacewings, bugs, ladybird beetles and spiders). Finally, cultural methods for control of leafhoppers can be supplemented as required by use of mild, contact insecticides such as fatty acid soaps, botanical pyrethrums and highly refined narrow-range oils. An important factor in the effectiveness of IPM as applied to leafhoppers, and to most other pests, is to monitor the size of the pest population. Simple surveys of nymph and adult populations can tell the grower if his strategies are succeeding or whether, as a last resort, it is necessary to apply conventional pesticides.

CONTROL OF SOIL-BORNE PESTS

The use of resistant rootstocks is likely to remain the central strategy for avoidance of damage by phylloxera and nematodes. Selection for resistance to soil-borne pests within the wild grapevines of North America has had a long and successful history (Chapter 7), and there are still large reserves of unexploited genetic variation in the genera *Vitis* and *Muscadinia* for use in rootstock breeding.

In the case of phylloxera there are no practical alternatives to resistant rootstocks. The rate of decline of susceptible vineyards can be slowed by avoidance of water stress, and flooding has been used to control phylloxera in the south of France. Generally, phylloxera cannot be controlled by cultural methods; chemical methods, fumigants or chemicals which move through the soil profile, are only partly successful because of lack of penetration. Similarly, the use of resistant rootstocks is the preferred method for avoiding nematode damage, but cultural techniques can lessen the impact of nematode infestations in susceptible vineyards. Any practice that enhances root growth and nutrient uptake is likely to reduce the effect of damage from the feeding of nematodes. The use of manures, to build up soil organic matter and to increase vine vigor, is often beneficial. Soil organic matter may have a regulating effect on nematode populations by providing habitats for parasitic fungi and bacteria. Cover crops with nematode-suppressant-species such as *Tagetes* (marigolds) are another technique for manipulation of the soil ecosystem for pest control.

Soil fumigation at the time of planting with methyl bromide or 1,3-dichloropropene are effective in the short term for nematode control, and are particularly valuable in replant situations where endoparasitic nematodes may survive in root fragments from the previous vineyard. The lack of complete penetration of fumigants into the soil leads to reinfestation of treated vineyards within a few years.

FUTURE DIRECTIONS IN PEST CONTROL

The advent of environmentalism and of increasing restrictions on use of chemical pesticides has led to interest in mechanical methods of pest control. Machines are being developed for vineyard use which act as vacuum cleaners; they sweep the foliage and collect and destroy the insect pests. The availability of such machines will add to the armory of environmentally sensitive control measures. Mechanical devices are likely to be most useful in reducing pest populations to manageable levels for control by other IPM techniques.

Biological methods of pest control are receiving great attention. The use of chemical signals to modify the behavior of insects and mites is an already well-developed field in plant protection. Attractants and repellents that are specific to various pest species of grapevines are either

available or are foreshadowed. Another well-established method of biological control is to disrupt the mating of target species by release of
sterile males and thereby achieve a reduction in population growth.

Spray materials containing entomotoxins such as BT (*Bacillus
thuringiensis* toxin) have been available for several years for control of
caterpillars. The next step is to incorporate genes for BT production
directly into the genomes of grapevine cultivars and to arrange for the
expression of these genes where and when required by use of appropriate
promoters. Another potential application of molecular genetics is to
confer pesticide resistance or tolerance on the predator species of economically important pests of grapevines.

Biotic stress: diseases

Vitis vinifera is subject to numerous diseases, some of which are of local
importance and some of which are serious problems in all grape growing
countries. The common diseases of *vinifera* grapes are listed in Table 6.8;
the most important of these will be discussed in the following section.

Table 6.8. *Common diseases of Vitis vinifera L.*

Type	Common name	Causal organism
Fungal diseases of leaves and fruit	Powdery mildew	*Uncinula necator*
	Downy mildew	*Plasmopara viticola*
	Botrytis bunch rot and blight	*Botrytis cinerea*
	Black rot	*Guignardia bidwellii*
	Phomopsis cane and leaf spot	*Phomopsis viticola*
	Anthracnose	*Elsinoë ampelina*
Fungal diseases of vascular system and roots	Eutypa dieback	*Eutypa lata*
	Esca	Unknown
	Armillaria root rot	*Armillaria mellea*
	Verticillium wilt	*Verticillium dahliae*
Bacterial diseases	Crown Gall	*Agrobacterium tumefaciens*
	Pierce's Disease	*Xylella fastidiosa*
	Bacterial Blight	*Xanthomonas ampelina*
Diseases caused by viruses and virus-like agents	Fanleaf degeneration	Grapevine fanleaf virus (GFLV)
	Leafroll	Unknown, possibly closterovirus
	Corky bark	Unknown
	Rupestris stem pitting	Unknown
Miscellaneous	Flavescence dorée	Unknown, possibly a mycoplasma-like organism

FUNGAL DISEASES

POWDERY MILDEW

Powdery mildew, or oidium, is one of the most widespread fungal diseases of grapevines. The causal organism is *Uncinula necator*, and it infects all green tissues of the grapevine. Mycelium grows on the surface of infected tissue and produces structures which penetrate the epidermal cells. The absorption of nutrients by the fungus leads to necrosis of the infected tissues. Chains of conidia arise from the mycelium, and this gives the leaf or other infected organs a characteristic dusty or powdery appearance. Cluster infection occurs at the time of bloom and results in poor fruit set and reduced yield. Infected berries tend to crack or split, and this predisposes them to attack by other fungi. Shoots attacked by powdery mildew during the growing season give rise to canes with characteristic red or brown lesions on the bark. *Uncinula necator* overwinters as hyphae within dormant latent buds or as cleistothecia on the surface of the vine or on debris on the vineyard floor.

Sulphur, in the form of a dust or a wettable powder, is the most widely used fungicide. It has both preventative and curative actions, and is effective in its vapor phase. The optimal temperature range for sulphur activity is 25–30 °C. Cultural factors can reduce the severity of powdery mildew infections and increase the effectiveness of chemical control. Good canopy management, with adequate air circulation and sun-exposure, is highly beneficial.

DOWNY MILDEW

Downy mildew, caused by the fungus *Plasmopara viticola*, is primarily a disease of grape growing regions with summer rainfall. Downy mildew is not normally a problem in Mediterranean climates. *Plasmopara* attacks all green parts of the vine, particularly the leaves. Severe infections cause defoliation and this leads to reduced sugar accumulation by the fruit and to loss of winter hardiness by latent buds. The characteristic lesions are yellow or brown 'oily' spots in the interveinal laminar tissues. Sporulation of the fungus occurs on the lower leaf surface, where it produces a dense white cotton-like growth. Infected shoot tips become curled and young berries become covered with the downy felt of fungus sporulation. The classical control method for downy mildew, which was developed in the 1880s, is to spray with Bordeaux Mixture, a suspension containing copper sulphate and hydrated lime. In recent years,

Bordeaux Mixture has been replaced by systemic organic fungicides such as fosetyl aluminum and phenylamides, and by cymoxanil, a non-systemic penetrating fungicide specific to mildew.

BOTRYTIS BUNCH ROT AND BLIGHT, NOBLE ROT

Botrytis bunch rot (syn. grey mould, gray mold) is a serious problem in all of the world's vineyards. This disease greatly reduces both yield and quality. There is loss of bunches through rotting of the rachis, loss of juice and rotting and desiccation of the fruit. In table grapes, botrytis causes losses in transit and storage. In wine grapes, botrytis bunch rot has detrimental effects on wine quality. Fungal enzymes convert glucose and fructose to glycerol and gluconic acid and promote phenol oxidation. The fungus also secretes polysaccharides, which interfere with the clarification of wine. Affected grapes produce wines with off flavors.

The causal organism of botrytis bunch rot and blight is *Botrytis cinerea*. This fungus is not specific to grapevines; it attacks many cultivated plants and can live as a saprophyte. In grapes, botrytis attacks both the vegetative and reproductive tissues. In spring, infections occur in the young shoots, in leaves and in the immature, unexpanded inflores-cences. In all cases, this leads to shrivelling and abscission. From véraison onward the fungus enters ripening berries directly through the epidermis or through wounds. Compact clusters in which the berries are tightly pressed together are especially susceptible to infection by botrytis. There is variation among cultivars in susceptibility to botrytis bunch rot but there is no true resistance in *vinifera* grapes, and chemical control methods are required. In Europe, fungicides (benzimadozoles or dicarboximides) are applied at fruit set, during early fruit growth, at véraison and three weeks before harvest.

The occurrence of severe outbreaks of botrytis bunch rot is closely correlated with the incidence of rainfall during the pre-harvest period, and cultural techniques are very important in lessening the impact of the disease. The use of trellising, shoot placement and leaf-removal from the base of the cane are all techniques which limit the growth of botrytis by increasing air circulation and light penetration into the canopy.

Under certain conditions the growth of *Botrytis cinerea* on ripe bunches is attenuated and it takes on a form known as 'noble rot' (pourriture noble). This infection accelerates water loss from the berries and it en-hances the sweetness and flavor of the juice. Wines made from fruit affected by noble rot are very sweet and liquorous. Examples of 'botrytized' wines include the sauternes of France (made from Semillon) and the German wines known as Auslese, Beerenauslese and Trockenbeerenauslese (made from Riesling). The typical weather conditions for occurrence of noble rot are warm, dry autumn days with misty mornings. However, there is a fine line between noble rot and bunch rot, and a shower of rain in the period immediately before harvest can cause the benign noble rot to change very rapidly into rampant bunch rot.

EUTYPA DIEBACK

This disease, which is also known as dying arm, involves the death of the arms which bear the fruiting canes, but it is progressive and leads to the death of the entire vine. Dying arm is an increasingly severe problem in European, American and Australian viticulture. The causal organism is the fungus *Eutypa lata* (syn. *E. armeniacae*). Symptoms of eutypa dieback include the loss of apical dominance by the latent buds of spurs and the production of numerous weak axillary shoots with cupped leaves. There is also discoloration of the xylem and loss of xylem function. Lesions in the wood first appear as dark-colored sectors, which extend from the bark to the pith. As the disease progresses a greater proportion of the cross-sectional area of the stem becomes discolored and non-functional. *Eutypa lata* is a wound parasite and it induces the formation of wound cankers, particularly in older vines. Eutypa dieback is seldom a problem in vines less than five to six years of age. Once infection has occurred the disease progresses slowly and there is usually vigorous regrowth from healthy tissues towards the base of the vine. The productive life of vines can be prolonged by cutting away diseased tissue and by re-forming new arms or cordons from vigorous, healthy canes. The treatment of pruning cuts or wounds with the fungicide benomyl is an effective control measure for this disease.

BACTERIAL DISEASES

PIERCE'S DISEASE (PD)

PD is a fatal disease of grapevines in all parts of North and Central America which have mild winters. In the southeastern United States, PD is the limiting factor to the growing of *vinifera* grapes. The symptoms of PD resemble those of water stress and include the drying or scorching of leaves. There is also irregular lignification and suberization of the canes and some or all of the fruit clusters may shrivel and die.

The causal agent of PD is a Gram-negative bacterium, *Xylella fastidiosa* (Davis *et al.*, 1978; Wells *et al.*, 1987). The PD bacterium is restricted to the xylem elements of the cane, where it induces formation of tyloses which, in turn, disrupt water movement. The bacterium has a wide host range including many grasses, brambles (*Rubus* spp.), elderberry (*Sambucus* spp.) and willows (*Salix* spp.). These hosts act as reservoirs for the PD bacterium. The vectors of PD are sharpshooters, a type of leaf hopper. In California, three species of sharpshooter, *Draeculocephala minerva* (green sharpshooter), *Graphocephala atropunctata* (blue-green sharpshooter) and *Carneocephala fulgida* (red-headed sharpshooter) are responsible for the spread of Pierce's disease. PD does not move from vine to vine. There is no secondary spread by sharpshooters or by other means. Currently, the most effective control is to prevent the sharpshooter vectors from entering the vineyard. This is best achieved by careful site selection and by removal or avoidance of alternative host plants. PD 'hot spots' often occur in vineyards downwind from riparian habitats or from weedy areas, and these circumstances must be avoided.

CROWN GALL

The causal organism of crown gall disease is *Agrobacterium tumefaciens*, a soil-inhabiting, Gram-negative bacterium. The bacterium enters the grapevine through wounds in the trunk or arms, and infection leads to the formation of fleshy galls. *Agrobacterium* contains an infective plasmid, a circular element of DNA. In the infection process, this plasmid DNA is inserted into the genome of the host cell. Information contained in the bacterial plasmid causes infected cells to overproduce the plant hormones auxin and cytokinin. This leads to cell division and to the formation of the characteristic tumors or galls. Other bacterial genes that are expressed in host cells cause the production of opines, amino-acid-like compounds which are metabolized only by *Agrobacterium*. There are three biovars of the pathogen, of which biovar 3 is predominant on grapevines. The use of *Agrobacterium tumefaciens* as a vehicle for transformation and genetic improvement of grapevines is discussed in greater detail in Chapter 7.

Agrobacterium tumefaciens is systemic within grapevines and can be spread by use of infected propagating materials. There is often a high incidence of gall formation on cuttings and bench-grafts. *Agrobacterium* infections are most damaging to young vines during vineyard establishment; rapidly growing galls can girdle young vines in a single season. A common treatment for older vines is to cut away the galls and treat wounds with kerosene. Because of the systemic nature of the pathogen, new galls are likely to be formed and chemical control is very difficult. In established vines, crown gall is seldom a fatal disease but heavily infected vines have reduced yield and vigor. The most effective method of control is to use disease-free planting materials, particularly when establishing vineyards in new areas. Biovar 3 has not been detected in soils in which grapes have not been grown previously.

VIRUS DISEASES AND DISEASES WITH VIRUS-LIKE SYMPTOMS

Grapevines, in common with all other woody perennial fruit plants, are subject to infection by viruses and virus-like organisms. The disease fanleaf degeneration is known to be caused by a nepovirus, but the causal agents of some other widespread diseases with virus-like symptoms have yet to be identified. Included are leafroll, corky bark and rupestris stem pitting.

The virus and virus-like diseases of grapevines are all transmissible by grafting; none is transmissible by seed or by pollen. The use of planting material that is free from all known pathogens is of paramount importance in viticulture. Obviously, grapevines with recognizable symptoms of disease are not used as sources of cuttings or budwood, but seemingly healthy mother plants can be symptomless carriers of diseases that have devastating effects on vineyard productivity.

Elaborate and time-consuming indexing procedures are necessary to identify clean propagating material. New diagnostic tools for virus disease in fruit crops include immunological methods (e.g., ELISA, enzyme-linked immunosorbent assay) and nucleic acid technology (ds-RNA, c-DNA probes), but conventional indexing using sensitive indicator plants is still required in circumstances where the causal agents of the disease are unknown.

There has been much research since the 1960s on the heat-inactivation of grapevine viruses and on the use of thermotherapy for production of healthy propagating material (Bovey and Martelli, 1986). Similarly, there has been much research on applications of tissue culture for phytosanitary processes. Included has been work on the exploitation of 'natural escape' by culture of isolated apices, and combination of shoot apical culture and thermotherapy (Galzy, 1964; Goheen and Luhn, 1973; Bass and Vuittenez, 1977).

In most grape growing countries, clean stock programs have been established, either under government control or government surveillance, to produce certified true-to-type, pathogen-free propagation stock for use by the nursery trade. For example, the Foundation Plant Materials Service of the University of California, Davis, which was the first program of its type, has been outstandingly successful in safeguarding the health of US vineyards. In the following section, brief descriptions will be given of each of the principal virus and virus-like diseases of grapevines.

FANLEAF DEGENERATION

Fanleaf degeneration, a disease caused by grapevine fanleaf virus (GFLV), is named for the fan-shaped leaves of infected vines. Other external symptoms include abnormal shoot morphology (double nodes, short internodes, fasciations), yellow mosaic (yellow discolorations of leaves, stems, tendrils and inflorescences), and veinbanding (yellowing along main veins of mature leaves). Grapevines infected with GFLV produce small straggly bunches in which fruit set and fruit ripening is irregular. The severity of fanleaf degeneration varies among cultivars. In sensitive genotypes (e.g. Cabernet Sauvignon) yields are reduced by 80% and the productive life of the vineyard is greatly reduced. Infected propagating material is often difficult to root and to graft.

GFLV is a nepovirus, which is transmitted from plant to plant by the dagger nematode, *Xiphinema index*. Long-range spread of GFLV is by use of infected planting material. The natural host range is restricted to the genus *Vitis*, but GFLV is transmissible to a wide range of herbaceous species by sap-rubbing inoculation. Useful diagnostic species (indicator plants) include *Chenopodium amaranticolor, Gomphrena globosa* and *Cucumis sativus*. Fanleaf degeneration is the only so called 'virus disease' of grapevines in which a virus particle is known to be the causal agent, and immunological methods can readily be applied to its detection. GFLV isolates are antigenically uniform and diagnosis by ELISA is a standard procedure. Strategies for control of fanleaf degeneration include (i) control of nematode vectors by long-term fallowing or by soil fumigation; (ii) breeding rootstocks for resistance to the feeding of *Xiphinema* nematodes; and (iii) breeding grapevines for resistance to GFLV virus. Fallowing is generally uneconomic as a nematode control measure, and the benefits of soil fumigation are temporary. Breeding is the most promising approach but it is a long-term prospect. Meanwhile, the spread of fanleaf degeneration can be limited by planting only certified disease-free grapevines.

LEAFROLL

Leafroll is found wherever grapes are grown. As its name implies, symptoms of leafroll include an inward rolling of leaves in late summer. The formation of intense, premature autumn colors is also a characteristic symptom. Leafroll is not a fatal condition but it causes substantial reductions in yield, vigor, and fruit quality. In affected vines, the ripening of fruit is delayed, and pigment synthesis and sugar accumulation are reduced. The identity of the causal agent of leafroll is not yet clear. Particles resembling closteroviruses are often found in diseased vines and there is also evidence of an association with a viroid. The natural spread of leafroll in vineyards is very slow and no vectors have been identified. The only important method of spread is the use of infected propagating material. Leafroll can be eliminated from nursery stock by indexing procedures. The cultivar Cabernet franc is a sensitive indicator for leafroll. Scions from candidate mother vines are grafted to Cabernet franc. If no symptoms are developed by the indicators after 18 months the candidate is free of disease and can be registered as a mother vine source. Propagation from registered mother vines is a highly efficient method of controlling the spread of leafroll. The application of modern immunological methods to the detection of systemic plant pathogens such as those responsible for leafroll will greatly speed up the indexing process.

CORKY BARK

Corky bark is another widespread disease of grapevines that is disseminated by use of infected propagating material, especially rootstocks. *Vitis rupestris* cv. St George, a widely used rootstock, is very susceptible to corky bark. Infected vines have a thick, roughened bark and secondary thickening is irregular. The trunks of vines infected with corky bark are deeply grooved in cross section, and the wood is pink in color. Other rootstocks which exhibit stem-grooving are Harmony, 1613 Couderc and 110 Richter.

Bark symptoms are less noticeable in *Vitis vinifera* cultivars than in many species or hybrids. Many *vinifera* cultivars that carry the disease do not show symptoms until grafted onto species or hybrid rootstocks. In these cases, an incompatibility develops at the graft union, which results in the eventual death of the scion. The highly susceptible interspecific hybrid LN-33 is used as an indicator of corky bark. In this genotype there are severe symptoms of abnormal cambial activity (bark swelling, stem-grooving) and the vines die soon after inoculation. The causal agent of corky bark is unknown and there are no known vectors. The disease is easily avoided by use of clean, certified, planting material.

RUPESTRIS STEM PITTING

This disease is regarded as important in California, where *Vitis rupestris* SCHEELE is a parent of one of the most widely-used rootstocks, AxR # 1 (syn.: ARG-1, Ganzin-1; *V. rupestris* × *V. vinifera*) and where there is a substantial area of vines grafted onto St George (syn. Du Lot), a selection of *Vitis rupestris*.

In St George, the indicator for rupestris stem pitting, characteristic pits are produced in the wood close to the point of inoculation. The disease is said to reduce vigor and yield, but there has been little experimental work on its viticultural significance. Rupestris stem pitting, the causal agent of which is unknown, is not recognized as an important disease in Europe or Australia.

Recommended reading

Bovey, R. 1982. Control of virus and virus-like diseases of grapevine: sanitary selection and certification, heat therapy, soil fumigation and performance of virus-tested material. In *Proc. 7th Meeting, International Council for the Study of Viruses and Virus Diseases of the Grapevine (ICVG), Niagara Falls, Canada* (ed. A.J. McGinnis), pp. 299–309. Vineland Research Station, Ontario, Canada.

Flaherty, D.L., Jensen, F.L., Kasimatis, A.N., Kido, H. and Moller, W.J. (eds). 1982. *Grape pest management.* Cooperative Extension, University of California, Division of Agriculture and Natural Resources. Publication 4105, Berkeley, California. 312 pp.

Pearson, R.C. and Goheen, A.C. (eds). 1988. *Compendium of grape diseases.* American Phytopathological Society Press, St. Paul, Minnesota. 93 pp.

Literature cited

Ahmedullah, M., Kawakami, A., Sandidge, III, C.R. and Wample, R.L. 1986. Effect of paclobutrazol on the vegetative growth, yield, quality and winter hardiness of buds of 'Concord' grape. *HortSci.* **21**: 273–4.

Alexander, D. McE. 1958. Seasonal fluctuations in the nitrogen content of the Sultana vine. *Aust. J. Agric. Res.* **8**: 162–78.

Araujo, F.J. 1988. *The response of three-year-old Thompson Seedless grapevines to drip and furrow irrigation in the San Joaquin Valley.* Masters thesis, University of California, Davis.

Archer, E. 1987. Effect of plant spacing on root distribution and some qualitative parameters of vines. In *Proc. 6th Australian Wine Industry Technical Conference* (ed. E.T. Lee), pp. 55–8. Australian Industrial Publishers, Adelaide, Australia.

Archer, E. and Strauss, H.C. 1985. Effect of plant density on root distribution of three-year-old grafted 99 Richter grapevines. *S. Afr. J. Enol. Vitic.* **6**: 25–30.

Archer, E. and Strauss, H.C. 1989. The effect of plant spacing on the water status of soil and grapevines. *S. Afr. J. Enol. Vitic.* **10**: 49–58.

Archer, E., Swanepoel, J.J. and Strauss, H.C. 1988. Effect of plant spacing and trellising systems on grapevine root distribution. In *The grapevine root and its environment* (ed. J.L. Van Zyl), pp. 74–87. Tech. Comm. 215, Dept. Agric. Water Supply, Pretoria.

Baldwin, J.G. 1964. The relation between weather and fruitfulness of the Sultana vine. *Aust. J. Agric. Res.* **15**: 920–8.

Balo, B., Mustardy, L.A., Hideg, E. and Faludi-Daniel, A. 1986. Studies on the effect of chilling on the photosynthesis of grapevine. *Vitis* **35**: 1–7.

Bass, P. and Vuittenez, A. 1977. Amélioration de la thermothérapie des vignes virosées au moyen de la culture d'apex sur milieux nutritifs ou par greffage sur vignes de semis obtenues aseptiquement *in vitro*. *Ann. Phytopathol.* 9: 539–40.

Bledsoe, A.M., Kliewer, W.M. and Marois, J.J. 1988. Effects of timing and severity of leaf removal on yield and fruit composition of Sauvignon blanc grape vines. *Am. J. Enol. Vitic.* 39: 49–54.

Boulton, R. 1980a. The relationships between total acidity, titratable acidity and pH in wine. *Am. J. Enol. Vitic.* 31: 76–80.

Boulton, R. 1980b. The general relationship between potassium, sodium and pH in grape juice and wine. *Am. J. Enol. Vitic.* 31: 182–6.

Bovey, R. and Martelli, G.P. 1986. The viruses and virus-like diseases of the grapevine. *Vitis* 25: 227–75.

Brar, S.S. and Bindra, A.S. 1986. Effect of plant density on vine growth, yield, fruit quality and nutrient status of Perlette grapevines. *Vitis* 25: 96–106.

Carbonneau, A., Casteran, P. and Leclair, P. 1978. Essai de determination en biologie de la plante entière, de relations essentielles entre le bioclimat naturel, la physiologie de la vigne et la composition du raisin. Méthodologie et premiers résultats sur les systèmes de conduite. *Ann. Amélior. Plantes* 28: 195–221.

Christensen, P. 1969. Seasonal changes and distribution of nutritional elements in Thompson Seedless grapevines. *Am. J. Enol. Vitic.* 20: 176–96.

Christensen, P. 1980. Timing of zinc foliar sprays. I. Effects of application intervals preceding and during the bloom and fruit-set stages. II. Effects of day vs. night application. *Am. J. Enol. Vitic.* 31: 53–9.

Christensen, P. Kasimatis, A.N. and Jensen, F.L. 1978. Grapevine nutrition and fertilization in the San Joaquin Valley. *Agric. Sci. Univ. California, Berkeley Div. Agric. Sci. Publ.* no. 4087, 40 pp.

Clingeleffer, P.R. 1984. Production and growth of minimal pruned Sultana vines. *Vitis* 23: 42–54.

Conradie, W.J. 1980. Seasonal uptake of nutrients by Chenin blanc in sand culture: I. Nitrogen. *S. Afr. J. Enol. Vitic.* 1: 59–65.

Conradie, W.J. 1981a. Seasonal uptake of nutrients by Chenin blanc in sand culture: II Phosphorus, potassium, calcium and magnesium. *S. Afr. J. Enol. Vitic.* 2: 7–13.

Conradie, W.J. 1981b. Nutrient consumption by Chenin blanc grown in sand culture and seasonal changes in the chemical composition of leaf blades and petioles. *S. Afr. J. Enol. Vitic.* 2: 15–18.

Conradie, W.J. 1986. Utilization of nitrogen by the grape-vine as affected by time of application and soil type. *S. Afr. J. Enol. Vitic.* 7: 76–83.

Conradie, W.J. and Saayman, D. 1989. Effects of long-term nitrogen, phosphorus, and potassium fertilization on Chenin blanc vines. II. Leaf analyses and grape composition. *Am. J. Enol. Vitic.* 40: 91–8.

Cook, J.A., Ward, W.R. and Wicks, A S. 1983. Phosphorus deficiency in California vineyards. *Calif. Agric.* 37: 16–18.

Corzo, P.E. 1982. Improving budburst in tropical vineyards. In *Grape and Wine Centennial Symposium Proceedings, June 18–21, 1980, Davis, CA* (ed. A.D. Webb), pp. 154–5. University of California, Davis.

Crippen, D.D. Jr. and Morrison, J.C. 1986. The effects of sun exposure on the compositional development of Cabernet Sauvignon berries. *Am. J. Enol. Vitic.* 37: 235–47.

Davis, M.J., Purcell, A.H. and Thompson, S.V. 1978. Pierce's Disease of grapevines: isolation of the causal bacterium. *Science* 199: 75–7.

Dokoozlian, N.K. 1990. *Light quantity and light quality within Vitis vinifera grapevine canopies and their relative influence on berry growth and composition.* Ph.D. thesis, University of California, Davis.

Downton, W.J.S. 1977. Photosynthesis in salt-stressed grapevines. *Aust. J. Plant Physiol.* **4**: 183–92.

Downton, W.J.S. and Loveys, B.R. 1978. Compositional changes during grape berry development in relation to abscissic acid and salinity. *Aust. J. Plant Physiol.* **5**: 415–23.

El-Zeftawi, B.M. 1982. Effects of ethephon on cluster loosening and berry composition of four wine grape cultivars. *J. Hort. Sci.* **57**: 457–63.

Galzy, R. 1964. Technique de la thermothérapie des viroses de la vigne. *Ann. Epiphyt.* **15**: 245–56.

Goheen, A.C. and Luhn, C.F. 1973. Heat inactivation of viruses of grapevines. *Riv. Patol. Veg.* **9** (Supplement): 287–9.

Goldberg, S.D., Rinot, M. and Karu N. 1971. Effect of trickle irrigation on distribution and utilization of soil moisture in a vineyard. *Soil Sci. Soc. Amer. Proc.* **35**: 127–30.

Grimes, D.W. and Williams, L.E. 1990. Irrigation effects on plant water relations and productivity of Thompson Seedless grapevines. *Crop Sci.* **30**: 255–60.

Gubler, W.D., Marois, J.J., Bledsoe, A.M. and Bettiga, L. 1987. Control of Botrytis bunch rot of grape with canopy management. *Plant Dis.* **71**: 599–601.

Hale, C.R. 1977. Relation between potassium and the malate and tartrate contents of grape berries. *Vitis* **16**: 9–19.

Harrell, D.C. and Williams, L.E. 1987. The influence of girdling and gibberellic acid application at fruitset on Ruby Seedless and Thompson Seedless grapes. *Am. J. Enol. Vitic.* **38**: 83–8.

Hart, J.W. 1988. *Light and plant growth* (Topics in plant physiology, no. 1). Unwin Hyman, London. 204 pp.

Hawker, J.S. and Walker, R.R. 1978. The effect of sodium chloride on the growth and fruiting of Cabernet Sauvignon vines. *Am. J. Enol. Vitic.* **29**: 172–6.

Hedberg, R.R. and Raison, J. 1982. The effect of vine spacing and trellising on yield and fruit quality of Shiraz grapevines. *Am. J. Enol. Vitic.* **33**: 20–30.

Hunter, D.M., Wiebe, J. and Bradt, O.A. 1985. Influence of spacing on fruit yields and quality of grape cultivars differing in vine vigor. *J. Amer. Soc. Hort. Sci.* **110**: 590–6.

Idso, S.B., Jackson, R.D. and Reginato, R.J. 1977. Remote sensing of crop yields. *Science* **196**: 19–25.

Idso, S.B., Jackson, R.D., Pinter, P.J. Jr., Reginato, R.J. and Hatfield, J.L. 1981. Normalizing the stress-degree-day parameter for environmental variability. *Agric. Meteor.* **24**: 45–55.

Intrieri, C., Silverstroni, O. and Poni, S. 1986. Preliminary experiments on paclobutrazol effects on potted vines (*V. vinifera* cv. 'Trebbiano'). *Acta Hort.* **179**: 589–92.

Jackson, D.I., Steans, G.F. and Hemmings, P.C. 1984. Vine response to increased node numbers. *Am. J. Enol. Vitic.* **35**: 161–163.

Jackson, R.D., Idso, S.B., Reginato, R.J. and Pinter, P.J. Jr. 1981. Canopy temperature as a crop water stress index. *Water Resour. Res.* **17**: 1113–38.

Jacob, H.E. 1950. Grapegrowing in California. *Cir. Calif. Agric. Ext. Serv.* **116**: 1–80.

Kliewer, W.M. 1982. Vineyard canopy management--a review. In *Grape and wine centennial symposium proceedings, June 18–21, 1980, Davis, CA*, (ed. A.D. Webbe), pp. 342–52. University of California, Davis.

Kliewer, W.M. and Smart, R.E. 1989. Canopy manipulation for optimizing vine microclimate, crop yield and composition of grapes. In *Manipulation of Fruiting* (ed. C.J. Wright), pp. 275–91. Butterworth, London.

Kriedemann, P.E., Torokalvy, E. and Smart, R.E. 1973. Natural occurrence and utilization of sunflecks by grapevine leaves. *Photosynthetica* **7**: 18–27.

Kuroi, I., Shiraishi, Y., and Imano, S. 1963. Studies on breaking the dormancy of grape vines. I. Effects of lime nitrogen treatment for shortening the rest period of glasshouse-grown grape vine. *J. Jap. Soc. Hort. Sci.* **332**: 175–80.

Lafon, J., Couillaud, P., Gay-Bellile, F. and Levy, J.F. 1965. Rythme de l'absorption minérale de la vigne au cours d'un cycle végétatif. *Vignes Vins* **140**: 17–21.

Lavee, S. 1987. Usefulness of growth regulators for controlling vine growth and improving grape quality in intensive vineyards. *Acta Hort.* **206**: 89–108.

Lavee, S. and Haskal, A. 1982. An integrated high density intensification system for table grapes. In *Grape and wine centennial symposium proceedings, June 18–21, 1980, Davis, CA*, (ed. A.D. Webb), pp. 390–8. University of California, Davis.

Lohnertz, O. 1988. Nährstoffelementaufnahme von reben im Verlauf eines Vegetationszyklus. *Mitt. Klosterneuburg* **38**: 124–9.

Maas, E.V. 1987. Salt tolerance of plants. In *Handbook of Plant Science in Agriculture*, vol. 2 (ed. C.R. Christie), pp. 57–75. CRC Press, Boca Raton, Florida.

Marocke, R., Balthazard, J. and Correge, G. 1976. Exportations en éléments fértilisants des principaux cépages cultivés en Alsace. *C.R. Acad. Agric. France* **62**: 420–9.

Marois, J.J., Bledsoe, A.M., Gubler, W.D. and Bostock, R.M. 1987. Effects of spray adjuvants on development of *Botrytis cinerea* on *Vitis vinifera* berries. *Phytopathol.* **77**: 1148–52.

May, P. 1965. Reducing inflorescence formation by shading individual Sultana buds. *Aust. J. Biol. Sci.* **18**: 463–73.

May, P. and Antcliff, A.J. 1963. The effect of shading on fruitfulness in the Sultana. *J. Hort. Sci.* **38**: 85–94.

McKenry, M.V. 1984. Grape root phenology relative to control of parasitic nematodes. *Am. J. Enol. Vitic.* **35**: 206–11.

Morris, J.R., Cawthon, D.L. and Fleming J.W. 1980. Effects of high rates of potassium fertilization on raw product quality and changes in pH and acidity during storage of 'Concord' grape juice. *Am. J. Enol. Vitic.* **31**: 323–8.

Olson, R.A and Kurtz, L.T. 1982. Crop nitrogen requirements, utilization, and fertilization. In Nitrogen in agricultural soils (ed. F.J. Stevensen), pp. 567–604 Agronomy Society America, Madison.

Peacock, W.L., Broadbent, F.E. and Christensen, L.P. 1982. Late-fall nitrogen application in vineyards is inefficient. *Calif Agric.* **36**: 22–3.

Peacock, W.L., Christensen, L.P. and Andris, H.L. 1987. Development of a drip irrigation schedule for average-canopy vineyards in the San Joaquin Valley. *Am. J. Enol. Vitic.* **38**: 113–19.

Peacock, W.L., Christensen, L.P. and Broadbent, F.E. 1989. Uptake, storage and utilization of soil-applied nitrogen by Thompson Seedless as affected by time of application. *Am. J. Enol. Vitic.* **39**: 16–20.

Peacock, W.L., Jensen, F., Else, J. and Leavitt, G. 1977a. The effects of girdling and ethephon treatments on fruit characteristics of Red Malaga. *Am. J. Enol. Vitic.* **28**: 228–30.

Peacock, W.L., Rolston, D.E., Aljibury, F.K., and Rauschkolb, R.S. 1977b. Evaluating drip, flood, and sprinkler irrigation of wine grapes. *Am. J. Enol. Vitic.* **28**: 193–5.

Phatak, S.C., Austin, M.E. and Mason, J.S. 1980. Ethephon as harvest-aid for muscadine grapes. *HortSci.* **15**: 267–8.

Pierquet, P., Stushnoff, C. and Burke, M.J. 1977. Low temperature exotherms in stem and bud tissues of *Vitis riparia*. *J. Am. Soc. Hort. Sci.* **102**: 54–5.

Raschke, K. 1979. Movements of stomata. In *Encyclopedia of plant physiology*, vol. 7 (ed. W. Haupt and M.E.F. Feinleib), pp. 383–441. (new series), Springer-Verlag, Berlin.

Reynolds, A G. 1988. Response of Okanagan Riesling vines to training system and simulated mechanical pruning. *Am. J. Enol. Vitic.* **39**: 205–12.

Reynolds, A.G. and Wardle, D.A. 1989. Effects of timing and severity of summer hedging on growth, yield, fruit composition, and canopy characteristics of de Chaunac. II. Yield and fruit composition. *Am. J. Enol. Vitic.* **40**: 299–308.

Saayman, D. and Van Huyssteen, L. 1980. Soil preparation studies: I. The effect of depth and method of soil preparation and of organic material on the performance of *Vitis vinifera* (var. Chenin blanc) on Hutton/Sterkspruit soil. *S. Afr. J. Enol. Vitic.* **1**: 107–21.

Sachs, R.M. and Weaver, R.J. 1968. Gibberellin and auxin-induced berry enlargement in *Vitis vinifera* L. *J. Hort. Sci.* **43**: 185–95.

Sauer, M.R. 1968. Effects of vine rootstocks on chloride concentration in 'Sultana' scions. *Vitis* **7**: 223–6.

Savage, S.D. and Sall, M.A. 1982. Botrytis bunch rot of grapes: The influence of selected cultural practices on infection under California conditions. *Plant Dis.* **67**: 771–4.

Savage, S.D. and Sall, M.A. 1984. Botrytis bunch rot of grapes: Influence of trellis type and canopy microclimate. *Phytopath.* **74**: 65–70.

Shaulis, N., Amberg, H. and Crowe, D. 1966. Response of Concord grapes to light, exposure and Geneva Double Curtain Training. *Proc. Amer. Soc. Hort. Sci.* **89**: 268–80.

Shaulis, N. and Kimball, K 1955. Effect of plant spacing on growth and yield of Concord grapes. *Proc. Amer. Soc. Hort. Sci.* **66**: 192–200.

Shaulis, N., Kimball, K. and Tompkins, J.P. 1953. The effect of trellis height and training systems on the growth and yield of Concord grapes under a controlled pruning severity. *Proc. Amer. Soc. Hort. Sci.* **62**: 221–7.

Shaulis, N. and May, P. 1971. Response of 'Sultana' vines to training on a divided canopy and to shoot crowding. *Am. J. Enol. Vitic.* **22**: 215–22.

Shaulis, N. and Robinson, W.B. 1953. The effect of season, pruning severity, and trellising on some chemical characteristics of Concord and Fredonia grape juice. *Proc. Amer. Soc. Hort. Sci.* **62**: 214–20.

Shulman, Y., Nir, G., Bazak, H., and Lavee, S. 1986. Grapevine girdling by morphactin in oil. *HortSci.* **21**: 999–1000.

Shulman, Y., Nir, G., Fanberstein, L. and Lavee, S. 1983. The effect of cyanamide on the release from dormancy of grapevine buds. *Scientia Hort.* **19**: 97–104.

Skinner, P.W., Cook, J.A. and Matthews, M.A. 1988. Responses of grapevines cvs. Chenin blanc and Chardonnay to phosphorus fertilizer applications under phosphorus-limited soil conditions. *Vitis* **27**: 95–109.

Skinner, P.W. and Matthews, M.A. 1989. Reproductive development in grape (*Vitis vinifera* L.) under phosphorus-limited conditions. *Sci. Hortic.* **38**: 49–60.

Smart, R.E. 1973. Sunlight interception by vineyards. *Am. J. Enol. Vitic.* **24**: 141–7.

Smart, R.E. 1974. Photosynthesis by grapevine canopies. *J. Appl. Ecol.* **11**: 997–1006.

Smart, R.E. 1985. Principles of grapevine canopy microclimate manipulation with implications for yield and quality. A review. *Am. J. Enol. Vitic.* **36**: 230–9.

Smart, R.E. and Coombe, B.G. 1983. Water relations of grapevines. In *Water deficit and plant growth* (ed. T.T. Kozlowski), vol. 7, pp. 137–96. Academic Press, New York.

Smart, R.E., Robinson, J.B., Due, G.R. and Brien, C.J. 1985*a*. Canopy microclimate modification for the cultivar Shiraz. I. Definition of canopy microclimate. *Vitis* **24**: 17–31.

Smart, R.E., Robinson, J.B., Due, G.R. and Brien, C.J. 1985*b*. Canopy microclimate modification for the cultivar Shiraz. II. Effects on must and wine composition. *Vitis* **24**: 119–28.

Smart, R.E., Turkington, C.R., and Evans, J.C. 1974. Grapevine response to furrow and trickle irrigation. *Am. J. Enol. Vitic.* **25**: 62–6.

Stergios, B.G. and Howell, G.S. 1977. Effects of defoliation, trellis height and cropping stress on the cold hardiness of Concord grapevines. *Am. J. Enol. Vitic.* **28**: 34–42.

Stewart, B.A. and Nielsen, D.R. 1990. *Irrigation of agricultural crops.* Monograph Series No. 30, ASA-(SSA-SSSA), Madison. 1218 pp.

Szyjewicz, E. and Kliewer, W.M. 1983. Influence of timing of ethephon application on yield and fruit composition of Chenin blanc grapevines. *Am. J. Enol. Vitic.* **34**: 53–6.

Szyjewicz, E., Rosner, N. and Kliewer, W.M. 1984. Ethephon [(2-Chloroethyl) phosphonic acid, Ethrel CEPA], in viticulture–A review. *Am. J. Enol. Vitic.* **35**: 117–23.

Thomas, C.S., Marois, J.J., and English, J.T. 1988. The effects of wind speed, temperature and relative humidity on development of aerial mycelium and conidia of *Botrytis cinerea* on grape. *Phytopathology* **78**: 260–65.

Van Zyl, J.L. 1984. Response of Colombard grapevines to irrigation as regards quality aspects and growth. *S. Afr. J. Enol. Vitic.* **5**: 19–28.

Van Zyl, J.L. 1986. Canopy temperature as a water stress indicator in vines. *S. Afr. J. Enol. Vitic.* **7**: 53–60.

Van Zyl, J.L. 1987. Diurnal variation in grapevine water stress as a function of changing soil water status and meteorological conditions. *S. Afr. J. Enol. Vitic.* **8**: 45–52.

Van Zyl, J.L. 1988. Response of grapevine roots to soil water regimes and irrigation systems. In *The grapevine root and its environment* (ed. J.L. van Zyl), pp. 30–43. (Comp.), Tech. Comm 215, Dept. Agric. Water Supply, Pretoria.

Van Zyl, J.L. and Van Huyssteen, L. 1980. Comparative studies on wine grapes on different trellising systems: I. Consumptive water use. *S. Afr. J. Enol. Vitic.* **1**: 7–14.

Weaver, R.J. 1972. Plant growth substances in agriculture. W.H. Freeman, San Francisco. 594 pp.

Weaver, R.J., Kasimatis, A.N., Johnson, J.O. and Vilas, N. 1984. Effect of trellis height and crossarm width and angle on yield of Thompson Seedless grapes. *Am. J. Enol. Vitic.* **35**: 94–6.

Weaver, R.J., Manivel, L. and Jensen, F.L. 1974. The effects of growth regulators, temperature, and drying on *Vitis vinifera* buds. *Vitis* **13**: 23–9.

Weaver, R.J. and McCune, S.B. 1960. Effects of overcropping Alicante Bouschet grapevines in relation to carbohydrate nutrition and development of the vine. *Proc. Amer. Soc. Hort. Sci.* **75**: 341–53.

Weaver, R.J. and Pool, R.M. 1971. Chemical thinning of grape clusters (*Vitis vinifera* L.). *Vitis* **10**: 201–9.

Weaver, R.J., Shindy, W. and Kliewer, W.M. 1969. Growth regulator induced movement of photosynthetic products into fruits of 'Black Corinth' grapes. *Plant Physiol.* **44**: 183–8.

Wells, J.M., Raju, B.C., Hung, H.Y, Weisburg, W.G., Mandelco-Paul, L. and Brenner, D.J. 1987. *Xylella fastidiosa* gen. nov. sp. nov.: gram-negative, xylem-limited, fastidious plant bacteria related to *Xanthomonas* spp. *Int. J. Syst. Bacteriol.* **37**: 136–43.

Williams, L.E. 1987. Growth of 'Thompson Seedless' grapevines. II. Nitrogen distribution. *J. Amer. Soc. Hort. Sci.* **112**: 330–3.

Williams, L.E. and Biscay, P.J. 1991. Partitioning of dry weight, nitrogen and potassium in Cabernet Sauvignon grapevines from anthesis until harvest. *Amer. J. Enol. Vitic.* **42**: 113–17.

Williams, L.E., Biscay, P.J. and Smith, R.J. 1987. Effect of interior canopy defoliation on berry composition and potassium distribution in Thompson Seedless grapevines. *Am. J. Enol. Vitic.* **38**: 287–92.

Williams, L.E., Biscay, P.J. and Smith, R.J. 1989. The effect of paclobutrazol injected into the soil on vegetative growth and yield of *Vitis vinifera* L., cv. Thompson Seedless. *J. Hort. Sci.* **64**: 625–31.

Williams, L.E. and Grimes, D.W. 1987. Modelling vine growth – development of a data set for a water balance subroutine. In *Proc. 6th Aust. Wine Ind. Tech. Conf., Adelaide, Australia, July 14–17, 1986* (ed. T. Lee), pp. 169–74. Australian Industrial Publishers, Adelaide.

Williams, L.E. and Matthews, M.A. 1990. Grapevine. In *Irrigation of agricultural crops* (ed. B.J. Stewart and D.R. Nielsen), pp. 1019–55. *Agronomy Monographs*, ASA-CSSA-SSSA, Madison, WI, no. 30.

Williams, L.E. and Smith, R.J. 1991. Partitioning of dry weight, nitrogen and potassium and root distribution of Cabernet Sauvignon grapevines grafted on three different rootstocks. *Amer. J. Enol. Vitic.* **42**: 118–22.

Winkler, A.J. 1929. The effect of dormant pruning on the carbohydrate metabolism of *Vitis vinifera*. *Hilgardia* **4**: 157–73.

Winkler, A.J. 1954. Effects of overcropping. *Am. J. Enol. Vitic.* **4**: 4–12.

Winkler, A.J. 1969. Effect of vine spacing in an unirrigated vineyard on vine physiology, production and wine quality. *Am. J. Enol. Vitic.* **20**: 7–15.

Winkler, A.J., Cook, J.A., Kliewer, W.M. and Lider, L.A. 1974. *General viticulture*, Second edition. Univ. California Press, Berkeley. 710 pp.

Ziv, M., Melamud, H., Bernstein, Z. and Lavee, S. 1981. Necrosis in grapevine buds (*Vitis vinifera* cv. Queen of Vineyard). II. Effect of gibberellic acid (GA₃) application. *Vitis* **20**: 105–14.

7
Genetic improvement of grapevines

Introduction

The breeding of woody perennial fruit plants such as grapevines presents formidable technical difficulties. Cultivated grapes are highly heterozygous outcrossers and they do not breed true from seed. Moreover, the characters which make a good cultivar are polygenic in their inheritance and are controlled by large numbers of genes of minor effect. Few traits of viticultural importance are controlled by single genes with dominant alleles. Grapevine cultivars represent highly subtle gene combinations; these combinations are disrupted by the sexual process. The probability of recombining genes in a hybrid so as to recreate the essential characters of a traditional cultivar is very low. Having raised progeny, the grapevine breeder is faced with the problem of selection criteria. Wine grapes, which account for 80% of viticultural production, are bred for quality rather than for yield, and there are no simple, quick objective tests of wine quality. Microvinification procedures are used as an aid to selection but the process is still slow and subjective. Grape breeding programs are expensive, long-term enterprises. A consequence of these difficulties is that grapevine cultivars of established merit have been perpetuated by vegetative propagation, a mode of reproduction in which the genetic apparatus of the mother plant is reproduced with high fidelity in the offspring.

Innovation in winegrowing, whether at genetic level or in vineyard management, is made difficult by the so-called genotype × environment interaction (G × E). Some cultivars perform well in a wide range of environments but others perform best in particular environments. In French viticulture, information on G × E has been accumulated for several centuries; present-day customs or regulations that determine 'l'encépagement,' the use of cultivars in a given region, reflect the collective experience of many generations of vignerons and wine drinkers.

The history of viticulture in America and Australia is short by comparison, but it is already well established that some traditional European cultivars react favorably with certain environments. An example is Cabernet Sauvignon in the Napa Valley (California) and at Coonawarra (South Australia). These American and Australian environments bear little resemblance to the environment of origin of Cabernet Sauvignon (Bordeaux) and illustrate the remarkable adaptability of this cultivar. It is noteworthy that some traditional cultivars are lacking in adaptability. Gamay, for example, does not seem to produce wines of good quality when grown outside its native Beaujolais. Many European authorities claim that the physical and chemical properties of the soil have an over-riding effect on wine quality (Fregoni, 1977; Branas, 1980) but this view is disputed by American (Kissler and Carlton, 1969), Australian (Rankine et al., 1971) and South African (Saayman, 1977) researchers.

Another disincentive to genetic innovation in wine growing is the conservatism of consumers, producers and middlemen. The perception of good quality in table wines is based on a narrow spectrum of flavors, aromas and other gustatory attributes of a more or less tangible nature. Essentially, good quality is equated with the wine-making qualities of a small group of European cultivars within the species Vitis vinifera. The genus Vitis has more than 60 interfertile species, but species other than vinifera produce wines with unfamiliar or unconventional flavors. These flavors are evident in the wines of interspecific hybrids, for example the 'foxy' wines of New York State (V. × labruscana). Complex hybrids involving V. vinifera, V. rupestris, V. berlandieri and several other North American species were produced in France as phylloxera-resistant cultivars. The wines of these 'producteurs directs' are coarse and have unfamiliar flavors, and they have not been widely accepted. It seems that to be successful a new wine cultivar must be very similar to, or perhaps indistinguishable from, the traditional cultivar that it is designed to replace. Conservatism is also reflected in the system of Appellation d'Origine Contrôlée (AOC) in France, and in similar legislation in other European wine growing countries. The laws of AOC specify which areas of land may be used for grape growing, the cultivars to be used, the maximum yield and the maximum and minimum alcohol content of the wines. On the one hand, AOC maintains quality and reinforces the image of the existing product. On the other hand, such legislation tends to stifle innovation.

To summarize, modern wine production is based primarily on traditional cultivars. The styles of wines produced by these cultivars enjoy a high level of consumer acceptance and are firmly entrenched. A high technology has been developed to grow the traditional cultivars, and this has become established by custom or by law. As a consequence of these special circumstances, *clonal selection*, the exploitation of genetic variation within traditional cultivars, has become the most widely used procedure for improvement of wine grapes.

Nevertheless, useful progress has been made in breeding new cultivars of wine grapes by both intraspecific hybridization (crossing *vinifera* cultivars) and interspecific hybridization. Table grapes and raisin grapes are not subject to the same constraints as wine grapes with regard to tradition and consumer acceptance, and conventional breeding programs have produced several outstanding new cultivars. It is with rootstocks that breeding and selection has had greatest impact on viticulture. Finally, biotechnology holds great promise as a means of inserting new characters into the genomes of traditional cultivars without changing the cultivars concerned in any of their other characters, including wine quality. These three approaches to grapevine improvement, clonal selection, hybridization and biotechnology will now be discussed in greater detail.

Clonal selection

By definition, a clone is a population of plants all members of which are the descendants by vegetative propagation of a single individual. Vegetative propagation enables the perpetuation of the mother plant genotype. There is no meiosis, no segregation and no recombination of genes. Vegetative reproduction involves only mitosis, and this ensures that the genetic information encoded in the DNA of vegetative progeny is the same as in the mother plant. In forestry, the clone-mother-plant is known as the *ortet* and the vegetative propagules derived from an ortet are referred to as *ramets*. This terminology is useful, but it has not been widely adopted in horticulture or viticulture.

Viticulture is founded on vegetative propagation and on so-called 'clonal cultivars'. In theory, all members of a clone are genetically identical, but it is common knowledge that there is variation within clonal cultivars of wine grapes. This raises questions as to (i) the legitimacy of the terms 'clone' and 'clonal' in a viticultural context and (ii) the origins of clonal variation in grapevine cultivars.

POLYCLONAL ORIGIN

According to Rives (1961) the traditional cultivars of wine grapes are not clones *sensu stricto*, but are 'polyclones', populations of plants derived from several closely related ortets or clone-mother-plants. It follows that recently bred cultivars of known history are more likely to conform to

06 GENETIC IMPROVEMENT

the classical definition of a clone than do the traditional cultivars of ancient origin.

The notion of polyclonal origin is attractive in historical terms. One can envisage that Neolithic farmers made special collections of wild grapevines that were similar in appearance. However, the theory of polyclonal origin must be reconciled with genetic principles. Because the wild progenitors of cultivated grapes were dioecious (male and female flowers on separate plants), and thus obligate outcrossers, individuals of *Vitis vinifera* are highly heterozygous: they possess dissimilar alleles at many gene loci. Just as with human beings, each gamete produced by a grapevine, be it pollen grain or egg cell, is the result of genetic exchange between the members of each chromosome pair and then the separation of the two individual chromosomes into different gametes. Each chromosome pair segregates independently of the others. This process results in a very thorough shuffling of parental genes into countless new combinations in the gametes. Just as the probability of human parents producing offspring identical to either parent or to each other is infinitesimal, so is it highly unlikely that the characteristics of, say, Pinot noir would be duplicated in several wild seedlings. Sibling seedlings would, however, be likely to share at least some characteristics and might even resemble each other morphologically. Such similar sibling seedlings might have been the clone-mother-vines of highly variable cultivars like Pinot noir.

At present, the polyclonal origin of cultivars is a theory for which there is little evidence, but new techniques in molecular genetics such as RFLP (restriction fragment length polymorphism) analysis have the potential to provide objective information on the extent of genetic heterogeneity within grapevine cultivars (Soller and Beckmann, 1983; Helentjaris *et al.*, 1985). It is predictable that this subject will receive much attention during the next few years.

MUTATION

Most of the leading cultivars of wine grapes are of ancient origin and are likely to have accumulated by now a substantial load of somatic mutations. The occurrence and selection of 'bud-sports' has been an important element in the genetic improvement of pome and stone fruits and in citrus. The occurrence of such bud-sports is less well documented in grapes, but conscious selection of superior mother vines has been in operation for several centuries. The ease with which *vinifera* grapes are propagated by cuttings is likely to have facilitated the perpetuation of superior selections.

Experience with all crop plants is that beneficial mutations affecting yield and quality are rare, and that the effects of mutation are generally deleterious. It is probable that selection against negative characteristics has been, and remains, an important element in the success of clonal selection in wine grapes.

Mutational events normally occur in all dividing cells and those that occur in a cell in the apical meristem will give rise to a mutant sector in an otherwise normal shoot. This mixture of normal and mutant tissue is known as a *chimera*. If a bud occurs in a mutant sector, the shoot (or bunch) that emerges will carry the mutation. Chimeras are of several types, but periclinal chimeras are of some interest in viticulture. In a periclinal chimera, mutation is restricted to derivatives of the LI histogenic layer, and such plants are composed of a mutant 'skin' enclosing a normal interior. An example of periclinal chimera is seen in the cultivars Pinot meunier and Pinot noir. Meunier resembles Pinot noir in essential characteristics, but it is distinguishable by diffuse white hairs on the apex and expanding leaves, with hairless sections and shoots that are phenotypically similar to Pinot noir. The partners in the chimera Pinot meunier/Pinot noir are readily separated when Meunier is propagated *in vitro* by the fragmented apex technique (Skene and Barlass, 1983). A form of mutation known as somaclonal variation occurs in plant materials that have been regenerated *in vitro* (Larkin and Scowcroft, 1981). In wine grapes, somaclonal variation may provide additional opportunities for clonal selection and this will be discussed further in a later section.

PATHOGENS

There is much debate on the relative importance of genetic and non-genetic factors in clonal variation in grapevines. One school of thought is that genetic differences within cultivars are minimal and that clonal variation is attributable primarily to differences among mother plants in the occurrence of pathogens.

It should be noted that the term 'clonal selection' has a somewhat different meaning in different countries. In Europe, the production and evaluation of pathogen-free stock is an integral part of the clonal selection process. In the US and Australia, clonal selection tends to be concerned with the identification of superior mother vines from among vines that have been indexed and are known to be pathogen-free. In other words, the genetic component of clonal variation has been the predominant interest in the US and Australia.

Controversy in regard to the relative importance of the genetic component and the pathogen component of clonal variation is being fueled by the recent discovery that viroids may be present in heat-treated 'virus-free' grapevines (Flores *et al.*, 1985). Included are viroids resembling hop stunt (Sano *et al.*, 1985), and citrus exocortis (Rezaian *et al.*, 1988) and other viroid-like RNAs (Semancik *et al.*, 1987). Grapevine viroids are believed to be transmitted through propagation (Szychowski *et al.*, 1988), but viroid-free grapevines can be produced by culture *in vitro* of shoot apices (Duran-Vila *et al.*, 1988).

It has been proposed by Pena-Iglesias and Vecino (1987) that one of the causal agents of grapevine leafroll is viroid in nature, and there is definitive evidence that grapevine yellow speckle, a disease occurring in Australia, is attributable to a viroid (Koltunow and Rezaian, 1988). However, no correlation has yet been established between the presence of the recently identified cryptic viroids or viroid-like RNAs and any major disease of grapevines. The viticultural significance of the viroids found in otherwise 'pathogen-free' grapevines is not yet clear, and the possibility that viroids contribute to clonal variation in wine grape cultivars cannot be ruled out.

As with mutation, the viruses and virus-like disease of grapevines have generally deleterious effects on growth and production. A belief within some sections of the wine industry that vines affected by mild leafroll produce wines of higher quality than vines of the more vigorously growing pathogen-free selections is not supported by research. Until there is firm evidence to the contrary, the notion of 'virus management' in viticulture should be regarded as dangerous heresy, and growers and nurserymen should use only certified clonal material for propagation.

Another criticism of clean stock programs is that the heat treatment and tissue culture process lead to damaging changes in the nature of the plant material. At present, there is no evidence to support the charge of genetic change, but it is well established that thermotherapy and tissue culture lead to certain temporary morphological changes in wine grapes, including changes in leaf shape and reversion to juvenile morphology (Valat and Rives, 1973; Mullins *et al.*, 1979; Grenan, 1982). These transient changes have been comprehensively researched and can be avoided by appropriate precautions (Grenan, 1984).

MISNAMING

Some of the so-called clonal variation in wine grapes is attributable to errors in cultivar identification, and this variation is interclonal in origin rather than intraclonal.

Many grapevine cultivars are difficult to differentiate on the basis of phenotypic characters, and difficulties in identification are compounded by extensive synonymy. The same cultivar may have different names depending on the region in which it is grown, and different cultivars may have the same name. The occurrence of three different 'Rieslings' in Australia, Rhine Riesling (Riesling), Clare Riesling (Crouchen) and Hunter River Riesling (Semillon), is indicative of the potential for confusion and for erroneous naming.

Classical ampelography (Galet, 1979), and other morphometric methods involving multiple descriptors and multivariate analysis, are useful for cultivar identification (Alleweldt and Dettweiler, 1986; Swanepoel and DeVilliers, 1987) but are of inadequate sensitivity for discrimination at the clonal level. Similarly, the main application of isozyme analysis is in identification of cultivars (Wolfe, 1976; Dal Belin Peruffo et al., 1981; Benin et al., 1988; Parfitt and Arulsekar, 1989). Enzyme polymorphism is generally inadequate for the purposes of differentiating among clones within cultivars. Currently, the most promising approach to the problems of cultivar and clone identity is the use of RFLP analysis, a highly sensitive method for detecting small differences in DNA sequence (Soller and Beckmann, 1983; Tanksley et al., 1989). The application of DNA fingerprinting to viticulture will enable significant advances not only in grapevine identification but also in breeding and evolutionary studies.

CONCLUSIONS

The existence of variation within cultivars of wine grapes is firmly established and so is the terminology of clonal variation. Many grapevine cultivars may not be clones within the strict meaning of the term, but this distinction is academic. The origins of clonal variation are still speculative, but the advent of molecular biology has provided some powerful new tools for studying the genetic structure and heterogeneity of grapevine cultivars. Accurate methods for confirming the identity of cultivars and clones will become available within a few years.

The identification of superior grapevines within cultivars for use as mother plants remains the most important and most difficult step in the clonal selection process. From a biometrical standpoint, clonal selection in wine grapes is among the most challenging forms of field experimentation. Random variation from three sources, vineyard, winery and taste panel, is likely to obscure all but the most pronounced effects of clone on wine style or wine quality. Moreover, replication over sites is highly desirable because of the environmental plasticity of the grapevine and the need to account for genotype × environment interactions. The costs of long-term, high-precision field experimentation and properly designed and executed winemaking trials are high, but the benefits of clonal selection within grapevine cultivars are substantial (Becker, 1977; Antcliff, 1973; Cirami *et al.*, 1985).

Grapevine breeding

HYBRIDIZATION

Breeding programs based on intraspecific hybridization, that is, crosses involving genotypes of *Vitis vinifera*, have been very successful for table grapes, both seeded and seedless, and there have been a few notable successes with wine grapes. Müller-Thurgau (probably Riesling × Sylvaner) was raised in Geisenheim in 1892 and is now widely grown in Germany. More recent work in both California and Australia has produced new cultivars of wine grapes, which are well adapted to dry, hot, irrigated conditions (Olmo, 1948; Antcliff, 1975).

The genera *Vitis* and *Muscadinia* contain vast reserves of useful genetic variation, including resistance to important pests and diseases, and the potential for grapevine improvement by interspecific hybridization is almost limitless (Table 7.1). Unfortunately, interspecific hybrid wine grapes have a very bad reputation and are not readily accepted by producers and consumers. The poor wine quality of the hybrid phylloxera-resistant cultivars bred in France in the late nineteenth and early twentieth centuries, the so-called 'producteurs directs', has led to strong prejudice against hybrids in general and has further strengthened preference for the traditional *vinifera* cultivars. The growing of hybrids is prohibited by law in Germany, and the area of 'producteurs directs' in France is now minuscule. Improved hybrid cultivars with acceptable wine quality are grown in New York state (USA) and in southern Canada (Ontario), where winters are usually too cold for *Vitis vinifera*, but the wines they produce are primarily for local consumption.

Given the many obstacles to breeding new cultivars of wine grapes it is reasonable to question whether the goal of combining pest and disease resistance with high wine quality is achievable. The answer is in the affirmative, and the lead has been given by German grape breeders. Interspecific hybrid cultivars have been bred which are free from undesirable flavor components of American *Vitis* species and which produce wines that are indistinguishable from *vinifera* wines (Alleweldt and Possingham, 1988).

Table 7.1. *Grapevine breeding: principal sources of resistance*

Character	Species
Abiotic Stress	
Winter hardiness	*V. amurensis, V. riparia*
Lime-chlorosis	*V. vinifera, V. berlandieri*
Salinity	*V. berlandieri*
Fungal Diseases	
Downy mildew	*V. riparia, V. rupestris, V. lincecumii, V. cinerea, V. berlandieri, V. labrusca, M. rotundifolia*
Powdery mildew	*V. aestivalis, V. cinerea, V. berlandieri, V. labrusca, M. rotundifolia*
Botrytis bunch rot	American species
Bacterial Diseases	
Crown gall	*V. amurensis, V. labrusca*
Pierce's disease	*V. caribaea, V. coriacea, V. simpsonii, M. rotundifolia*
Nematodes	
Meloidogyne spp.	*V. champinii, M. rotundifolia, V. longii*
Xiphinema spp.	*V. rufotomentosa, M. rotundifolia*
Insects	
Phylloxera	*V. riparia, V. rupestris, V. berlandieri, V. cinerea, V. champinii, M. rotundifolia*

Modified from Alleweldt and Possingham (1988).

It is likely that changing consumer attitudes to use of chemicals in crop protection will lead to renewed interest in hybrid wine grapes that are resistant to pests and diseases, and to greater acceptance of new hybrid cultivars. The present status of grapevine breeding is summarized in Fig. 7.1.

GENETIC ASPECTS OF GRAPEVINE BREEDING

Most of the characters of grapevines that are significant in viticulture are each controlled by many genes of minor effect. Experience in the breeding of annual crops is that recurrent selection procedures are the most efficient for improvement of quantitative characters because favorable gene combinations can be concentrated within a population over a succession of generations. However, in breeding grapevines and other woody perennials, the problems of juvenility and long generation times, and the long time required for evaluation of seedlings, make it very

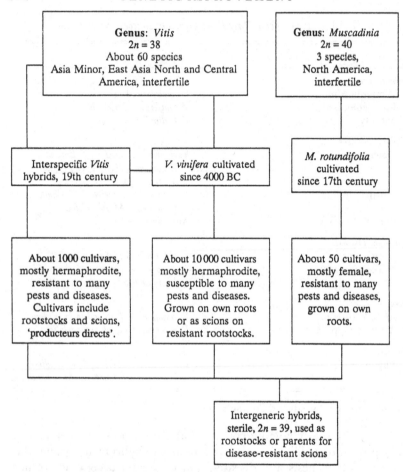

Fig. 7.1. Summary: history and present status of grapevine breeding. Modified from Alleweldt and Possingham (1988)

difficult to apply classical recurrent selection to the creation of new cultivars. Another constraint is the small size of the breeding population. Grape improvement programs are costly, and economic considerations require that population sizes be kept to a minimum. This is exemplified by the experience of Wagner and Bronner (1974) that the evaluation of a single seedling in a grape breeding program ties up 3–4 m² of prime viticultural land for at least seven years.

Schemes for grape breeding involving recurrent selection have been proposed (Bouquet, 1977), but the greatest progress has been achieved with the more 'traditional' methods for improvement of clonal cultivars.

These involve the crossing of heterozygous parents of different cultivars or species, and selection in the F_1 or later generations for one or a few of the best segregates. These selections are vegetatively propagated and tested for yield and quality in field trials. This latter step is especially important in breeding wine grapes because of the occurrence of strong genotype × environment interactions. Genetically speaking, the breeder seeks to make combinations with high general combining ability by selecting parents on phenotype and progeny performance. At the same time, the breeder hopes that specific combining ability, a reflection of non-additive genetic variance, will be in the positive direction.

Experience has shown that grapevines are improved by outcrossing and that the main cultivars are highly heterozygous. Most of the long-established cultivars carry a heavy load of deleterious recessive genes, and they exhibit strong inbreeding depression. Heterozygosity must be maintained in the breeding program by crossing the best representatives of unrelated lines.

The efficiency of grape breeding depends on the availability of suitable screening methods for fruit quality or wine quality, yield, disease resistance, winter hardiness, tolerance to chlorosis or salinity, and a range of other characteristics. Pre-selection criteria, that is, characteristics of seedlings which are indicative of the performance of adult plants, are of special importance in the breeding of woody perennials because much unwanted material can be eliminated at an early stage in the program. This greatly improves efficiency at later stages, which involve expensive field trials. In grapes, selection at the seedling stage for resistance to various pests and diseases has been successful but there are still no useful predictors of yield or quality. For the future, the application of biotechnology to grape breeding holds the promise of improved efficiency, especially through RFLP mapping and the identification of genes of viticultural significance (Helentjaris, 1988; Tanksley *et al.*, 1989).

BREEDING FOR DISEASE RESISTANCE

The inheritance of resistance to both downy mildew (*Plasmopara viticola*) and powdery mildew (*Uncinula necator*) has been studied by Boubals (1959, 1961) in *Vitis* and its relatives. Resistance to *Plasmopara* is expressed at two levels: (i) hypersensitivity at the point of infection (substomatal cavity) and (ii) inhibition of the subsequent growth of mycelium beyond the substomatal cavity. The hypersensitive reaction is simply inherited, but the second level of resistance is under polygenic control. There are also two levels of resistance to powdery mildew: (i) necrosis of appressoria within epidermal cells of resistance genotypes and (ii) necrosis of host cells following establishment of haustoria by the fungus. Inheritance of resistance to *Uncinula necator* is complex in *Vitis* but more simple in backcross progenies from *Vitis* × *Muscadinia* hybrids (Bouquet, 1986).

There is considerable variation among cultivars in susceptibility but there is no evidence of specific resistance to *Botrytis* in grapevines, and the breeder must rely on plant characteristics which confer an acceptable level of tolerance to the pathogen. These characteristics include lax bunches, which permit air circulation around the ripe berries, epidermis with thick cell walls and copious epicuticular wax, and elevated levels of phytoalexins such as resveratrol (Langcake and Pryce, 1976). Each of these plant characteristics is controlled by many genes and the availability of new cultivars with resistance to *Botrytis* is a very long-term prospect. The principal sources of pest and disease resistance in grapevine breeding are summarized in Table 7.1.

Rootstock breeding

The use of phylloxera-resistant rootstocks is one of the first (and best) examples of biological control of a pest of economic importance, and the grafting of susceptible *vinifera* scions onto rootstocks of resistant hybrids is still the foundation of viticulture. The selection and breeding of grapevine rootstocks has been in progress for more than 100 years. The first step was the collection of wild grapevines from the Mississippi Valley of the United States, where phylloxera is endemic. The root systems of these grapevines (pure species of *Vitis* or spontaneous interspecific hybrids) have a high degree of resistance to phylloxera, but much of the material from these early collections was unsuitable for use as rootstocks because it was difficult to propagate. Exceptions were a few genotypes of *Vitis rupestris* and *Vitis riparia*, which combined a high degree of phylloxera resistance with ease of propagation, graft compatibility and desirable effects on the yield and quality of *vinifera* scions.

Among the North American species, only *Vitis berlandieri* was well adapted to the highly calcareous soils of the premium grape-growing regions of France. This species, which occurs on calcareous outcrops in central Texas, is highly resistant to lime-induced chlorosis and to infestation by phylloxera, but *V. berlandieri* is unusable as a rootstock in its pure form because it is difficult to propagate by cuttings and grafting. These difficulties were overcome by interspecific hybridization with species that are easy to root and easy to graft. Most of the grapevine rootstocks in use today in French viticulture are hybrids derived from crosses with *V. berlandieri*. Another American species that was widely propagated in France at the turn of the century was *Vitis longii* (also known as *V. solonis*). This species was soon discarded as a phylloxera-resistant rootstock, but it was found later to be highly resistant to rootknot nematodes.

NEEDS AND PROSPECTS IN ROOTSTOCK BREEDING

None of the rootstocks presently available has all of the desirable characters required by viticulturists and winemakers, and all rootstocks have at least one serious weakness. Deficiencies include inadequate resistance to phylloxera and nematodes, inadequate tolerance to salinity, lime-induced chlorosis and water stress, and undesirable effects on the yield and quality of scions such as excessive vigor or inability to supply specific nutrient ions.

In the past, the positive attributes of many potential rootstock genotypes were negated at the level of the nursery because the mother plants were weakly growing, susceptible to pests and diseases of the foliage, and produced low yields of hardwood cuttings. In some cases these problems were compounded by the inherently low capacity for adventitious root formation of many North American species of *Vitis*, and by graft incompatibility with some cultivars or clones of *Vitis vinifera*.

There is great variation in the climatic, edaphic, biotic and economic conditions among the grape growing regions of the world, and it is impossible to write a prescription for the ideal rootstock, a conclusion reinforced by the importance of the genotype × environment interaction in viticulture. There is need for a wide range of rootstocks with differing attributes and it is fortunate that the genus *Vitis* contains a large reservoir of unexploited genetic variation. The prospects for breeding improved rootstocks for *vinifera* grapes are highly favorable.

The mechanisms of resistance at the field level for grapevine rootstocks to soil-borne pests should be defined before further discussion as there is discrepancy among viticulturists, entomologists and nematologists as to specific terminology. The degree of vine damage and soil pest build-up must be considered together if one is to develop working definitions of resistances for rootstocks used in phylloxera and nematode infested soils. The terminology of pest/plant interactions used in this book are as follows: (i) Susceptible response: vines damaged by the pest and responsible for high population increases; (ii) Tolerant response: vines which grow well in the presence of the pest but permit pest build-up; (iii) Resistant response: vines which grow well in the presence of the pest but greatly limit pest populations; and (iv) Immune response: soil pest never feeds and in fact avoids the vine.

PHYLLOXERA RESISTANCE

The so-called 'resistance' to phylloxera reported by many for grapevine rootstocks and wild species of *Vitis* is more accurately described as tolerance. Most of the phylloxera 'resistant' rootstocks used in Europe and elsewhere produce nodosities on their roots and, occasionally, galls on their leaves, when grown in infested soils, but they are able to outgrow the damage. The use of these tolerant rootstocks contributes to the perpetuation and spread of phylloxera, but the availability of genuinely resistant or immune rootstocks would lead to the eradication of phylloxera, the most insidious pest in the long history of viticulture. Breeding for resistance or immunity enables the rapid screening of large progenies because selections are made on an 'all-or-nothing' basis, but breeding for tolerance involves long-term trials of field resistance. A further disadvantage of tolerant rootstocks is the risk of breakdown of field resistance with the appearance of more aggressive biotypes of phylloxera.

Interest in breeding for true resistance or immunity has been stimulated by the finding that *Vitis cinerea* is a resistant species. This species was discarded by early rootstock breeders because of its poor rooting ability and susceptibility to lime-induced chlorosis, but the potential of *cinerea* germplasm has been reassessed (Zimmermann and Becker, 1978; Becker, 1988). Muscadine grapes (*Muscadinia rotundifolia*) also have a high degree of resistance or near-immunity to phylloxera, and this character exhibits dominance in crosses with *V. vinifera* (Davidis and Olmo, 1964; Bouquet, 1983).

NEMATODE RESISTANCE

Several rootstocks with resistance to the root knot nematode, *Meloidogyne incognita*, have been bred in California using germplasm from *Vitis champinii* and *Vitis longii*. This resistance is probably monogenic (Lider, 1954). Both of these species are susceptible to some races of *M. incognita*. The Californian rootstock Ramsey (Syn. Salt Creek, *V. champinii*), was introduced into Australia in the 1960s where it exhibits a high degree of resistance to *Meloidogyne javanica*. In France the main nematode pests are *M. arenaria* and *M. hapla*, and the common rootstocks vary from highly susceptible to resistant (Bouquet and Dalmasso, 1976). The basis of resistance to these two species of nematode is different because genotypes resistant to one are not necessarily resistant to the other. *M. hapla* is known to have races of differing virulence.

The muscadine grape is a promising source of germplasm for rootstock breeding because it exhibits a resistant response to *M. incognita, M. arenaria* and *M. javanica*. This resistance confers a high degree of dominance in its transmission, and it is unaffected by race-differences. Seedlings with resistance to both phylloxera and rootknot nematodes have been selected from advanced generations of V. *vinifera* × *M. rotundifolia* hybrids (Firoozabady and Olmo, 1982).

The dagger nematode, *Xiphinema index*, can be one of the most damaging pests of grapevines. It acts directly as a feeder on grapevine roots and indirectly as a vector of grapevine fanleaf virus (GFLV). The only available control measures for *X. index*, fallowing and soil fumigation, are unattractive to grape growers and provide only temporary relief. The use of immune or highly resistant rootstocks would prevent or reduce the reinfestation of healthy replants but immunity to this nematode species has not been found in *Vitis* or in related genera. All of the commercially available grapevine rootstocks are a host to *X. index* (Boubals and Pistre, 1978). Among *Vitis* species a useful level of resistance is shown by *V. rufotomentosa* (Kunde *et al.*, 1968; Harris, 1983). This resistance is probably monogenic, and it is dominant in transmission (Meredith *et al.*, 1982). Several other *Vitis* species are tolerant to *X. index* but tolerance is of no great viticultural interest because transfer of GFLV from the vector to the host requires only a few minutes of feeding, and even highly tolerant rootstocks are readily infected with the fanleaf virus. *Xiphinema index* is unable to feed on or colonize the roots of muscadine grapes (Boubals and Pistre, 1978), and it has been confirmed that *M. rotundifolia* has a high degree of resistance to GFLV transmission by *X. index* (Bouquet, 1981). Resistance to GFLV is present in some cultivars and F_1 hybrids of *M. rotundifolia* (Walker *et al.*, 1985). The development of the enzyme-linked immunosorbent assay (ELISA) for fanleaf virus enables the rapid screening of plants inoculated with viruliferous *X. index* (Bouquet and Danglot, 1983). The use of ELISA in combination with *Vitis* × *Muscadinia* hybridization holds considerable promise for the biological control of the *X. index* GFLV complex (Lider and Goheen, 1986).

RESISTANCE TO LIME-INDUCED CHLOROSIS
(IRON-CHLOROSIS)

Among the premier viticultural regions of France, Burgundy and Champagne both have highly calcareous soils, and the search for chlorosis-resistant rootstocks has been a preoccupation of French viticultural research. An important advance was the release of Fercal

(Pouget, 1980); this breeding program not only provided an improved rootstock cultivar but also improved biometrical strategies for rootstock breeding (Pouget and Ottenwaelter, 1973; Lefort and Leglise, 1977). A problem that remains is the unacceptably high vigor conferred upon the scion by many chlorosis-resistant rootstocks (Pouget, 1977).

SALT-TOLERANT ROOTSTOCKS

The grapevine is highly susceptible to salt damage; the breeding of chloride-excluding rootstocks is a practical means by which the productivity of salt-affected viticulture could be improved (Sauer, 1968; Bernstein et al., 1969). There is considerable genetic variation in salt-tolerance within the genus *Vitis* (Antcliff et al., 1983). In *Vitis berlandieri* there is evidence that chloride exclusion is governed by a single dominant gene (Newman and Antcliff, 1983).

ADAPTABILITY TO ACID SOILS

Ion-toxicity in acid soils is an increasing problem in viticulture, as it is in many other spheres of crop production. There is little genetic variation within the genus *Vitis* in resistance to high concentrations of copper, aluminum and manganese, the toxic ions which become increasingly available with decreasing pH. Some progress has been made in breeding a new rootstock for use in the acid gravel-based soils of Bordeaux (Pouget and Ottenwaelter, 1986), but the correction of soil pH by liming is probably the best solution to the acidity problem in most other grape growing regions.

MAGNESIUM DEFICIENCY

Many currently available rootstocks lack the capacity to supply adequate Mg^{2+} to the scion when grown on soils fertilized with high levels of potassium, or when used with cultivars that have high requirements for magnesium. Selection of rootstocks with resistance to magnesium deficiency is a difficult task because of stock–scion interaction. The absorption ability of the root system is influenced by the ability of the petiole to translocate Mg^{2+} and by the ability of the lamina to accumulate Mg^{2+}. Pouget and Delas (1982) used a technique of reciprocal grafting of stock and scion to clarify these interactions, and they were able to rank rootstock cultivars with respect to their abilities to absorb magnesium. This technique has great merit for the screening of new genotypes.

DROUGHT TOLERANCE

Several widely-used rootstocks possess a good level of tolerance to water stress, particularly hybrids of *V. rupestris* and *V. berlandieri*, but their usefulness for high-quality wine production is limited by the excessive vigor they impart to the scion under favorable growing conditions. In rootstock breeding, drought tolerance can be predicted by the responses of seedlings to water stress as indicated by measurements of stomatal resistance and leaf water potential (Fregoni *et al.*, 1978; Carbonneau, 1985).

Application of biotechnology to the genetic improvement of grapevines

The advent of tissue culture and genetic engineering, and the application of these technologies to crop improvement, has much significance for viticulture. There are two main areas of interest: (i) procedures that improve the efficiency of conventional breeding; and (ii) applications of cell and tissue culture that augment genetic variation within existing genotypes, i.e. clonal variation. The former has a major role in the breeding of rootstocks, table grapes and raisin grapes. The latter is of particular importance for premium wine grapes because of the 'genetic straitjacket' within which this form of viticulture is constrained by tradition, legislation and the market place.

PROGRESS IN GRAPEVINE TISSUE CULTURE

A prerequisite for the application of tissue culture to grapevine improvement is the availability of highly efficient methods for plant regeneration or plant propagation *in vitro*. The grapevine was among the first plants to be cultured *in vitro* (Morel, 1944). Proliferation of callus and formation of adventitious roots were the subjects of several reports during the 1950s and 1960s, but the grapevine proved to be recalcitrant with respect to regeneration *in vitro*. Somatic embryogenesis was reported by Mullins and Srinivasan in 1976, and organogenesis was reported in the same year by both Favre (1976) and Hirabayashi *et al.* (1976). These developments were pre-dated by the first report on cultivation of grapevine protoplasts (Skene, 1974). Since then, there has been much research on methodological factors affecting isolation, survival and division of grapevine protoplasts, but plant regeneration has yet to be achieved.

There has been extensive research on micropropagation of grapevines since the original report of Jona and Webb (1978), and highly efficient methods are available for induction of axillary shoot proliferation and subsequent formation of adventitious roots by microcuttings. A novel method for rapid multiplication *in vitro* using fragmented shoot apices was developed by Barlass and Skene (1978). In this procedure, adventitious buds are formed with very high frequency in the tissues produced by cultured leaf primordia. Recently, it has been shown that numerous adventitious buds can be induced on hypocotyl explants of somatic embryos of grape cultivars (Vilaplana and Mullins, 1989).

There is a single report from China of haploid plantlet production in grapevines (Zou and Li, 1981), but attempts elsewhere to obtain haploids by culture of anthers and pollen of *Vitis vinifera* have been unsuccessful. In many grapevine genotypes the connective of anthers is a highly regenerative tissue and gives rise to somatic embryos with high frequency. Callus produced by cultured anthers may contain haploid metaphases or nuclei in which the DNA content is consistent with the haploid condition, but derivatives of these cells do not seem to participate in embryo formation; plants from anther callus are diploid and heterozygous (Rajasekaran and Mullins, 1983b). Classical androgenesis involving internal divisions in pollen grains and extrusion of embryogenic callus, as seen in many *Solanaceae* and *Cruciferae*, has not been observed in grapevines. At the level of the intact plant, mixoploidy was observed in twinned seeds by Bouquet (1982), but no haploid individuals were recovered.

Application of the embryo-rescue technique to stenospermocarpic grapes has enabled 'seedless–seedless' hybridization. With this technique 'seedless' genotypes can be used as both male and female parents because zygotic embryos are rescued before they abort. This greatly increases the frequency of seedless progeny (Ramming and Emershad, 1982; Cain *et al.*, 1983).

PRACTICAL APPLICATIONS

So far, seedless–seedless hybridization has had the greatest impact of any aseptic method on grapevine improvement, and it represents a major improvement in the methodology for breeding seedless table grapes. Haploids, and homozygous diploids derived from them, would be particularly useful for grapevine breeding and for genetic studies. Haploid grapevines are still unavailable; it is possible that haploidy may be a lethal condition in the clonal cultivars of *Vitis vinifera* L.

The main application of micropropagation has been in the production of pathogen-free stock. Tissue culture was first used for virus elimination in the 1960s (Galzy, 1964) and is now a standard procedure in clean stock programs. Recently, the fragmented apex technique has been used to produce grapevines that are free from infection by viroids (Duran-Vila *et al.*, 1988). The relative ease with which nodal explants of grapevine cultivars can be induced to proliferate axillary buds has led to the use of micropropagation as a vehicle for mutation breeding (Reisch *et al.*, 1985; Barlass, 1986; Kim *et al.*, 1986). However, the usefulness of induced mutation for grape cultivar improvement has yet to be established.

In terms of potential applications there is much interest in the possibility that tissue culture procedures may be used to create or amplify genetic variation within commercially-important cultivars of wine grapes and, thereby, provide new raw material for clonal selection. These potential applications are founded on the processes of somatic embryogenesis, organogenesis, and on the exploitation of both random genetic variation and directed genetic change.

SOMACLONAL VARIATION IN WINE GRAPES

The random spontaneous genetic variation that arises during plantlet formation *in vitro* is termed 'somaclonal'. Variations of sugar cane, potato, rice, wheat, barley and rape have been discovered which possess disease resistance and several other agronomically interesting characters, and there has been enthusiastic speculation on the potential of somaclonal variation in perennial plant breeding (De Wald and Moore, 1987). In viticulture, there is much interest in somaclonal variation because it could provide a means of augmenting clonal variation within the traditional cultivars.

In addition to variation that arises as a consequence of tissue culture procedures, there are other potential sources of genetic variation in long-established cultivars of vegetatively propagated plants which are dependent upon tissue culture for their expression. Many fruit cultivars that arose as somatic mutations are chimeric in structure, and rearrangements in chimeric structure occur during plant regeneration *in vitro* (Skene and Barlass, 1983). In addition, ancient clones such as the traditional cultivars of grapevines are likely to have accumulated a considerable load of mutations over the centuries, and cell culture methods may provide a means by which this normally covert variation can be expressed.

Grapevines of most major cultivars and many hybrids have now been regenerated *in vitro* by somatic embryogenesis using nucellar tissues of unfertilized ovules (Mullins and Srinivasan, 1976) or the vegetative tissues of anthers (Rajasekaran and Mullins, 1983*b*). Hundreds, if not thousands, of grapevines have been produced from somatic embryos by researchers in several countries. Evidence of somaclonal variation has come primarily from research on genotypes that are highly regenerative *in vitro*, for example Gloryvine, a *Vitis vinifera* × *Vitis rupestris* hybrid. Gloryvines raised from somatic embryos often exhibit abnormalities such as dwarfism and albinism. Leaf shape is normally a highly stable character in grapes and is the basis of ampelography, but plants produced *in vitro* often show marked variations in leaf shape, including differences in petiolar sinuses and lobation. These differences tend to be transient and may be similar in nature to the temporary variations which occur in thermotherapy (Valat and Rives, 1973) or after micropropagation *in vitro* (Cancellier and Cossio, 1988). In addition, Gloryvines raised from somatic embryos show variation in sex expression (Rajasekaran and Mullins, 1983*a*), indicating, perhaps, change in a single gene (Antcliff, 1980). *Eutypa lata* is a toxin-producer (Mauro *et al.*, 1988); attempts have been made to select for resistance *in vitro* to the toxin in cell suspension culture and then to regenerate somaclonal variant grapevines by somatic embryogenesis which are resistant to Eutypa dieback.

In research on selection for salinity tolerance in *Vitis rupestris* SCHEELE cv. St George (Lebrun *et al.*, 1985), cell lines were selected which grew in suspension cultures containing up to 150 mM NaCl. These apparently salt-tolerant cell suspensions gave rise to somatic embryos, but the embryos became necrotic and died in the presence of 50 mM NaCl once radicle elongation had commenced. From these results it appears that somaclonal variation in NaCl-tolerance is manifested by cell suspensions of St George but that tolerance at cellular level and in immature embryos is not closely correlated with tolerance in fully differentiated embryos and in intact plants. So far, studies on somaclonal variation in wine grapes have given equivocal or disappointing results, but it is still premature to conclude that this source of variation has nothing to offer to grapevine improvement.

The variation that arises in tissue culture, or which is induced by mutagens, is essentially random in nature; its successful exploitation is dependent upon the availability of rapid, accurate screening procedures. The development of these methods is relatively straightforward for characters such as disease resistance. Micropathogenicity tests are already available for *in vitro* selection for resistance to downy mildew (*Plasmopara viticola*) (Morel, 1948; Lee and Wicks, 1982) and powdery mildew (*Uncinula necator*) (Klempka *et al.*, 1984) and selection at the level of phytoalexin production is an interesting possibility (Stein and

Hoos, 1984). It must be emphasized, however, that selection among somaclones for qualitative characters such as wine quality will remain as difficult and as time-consuming as conventional clonal selection with conventionally propagated grapevines.

PROTOPLAST TECHNOLOGY

The role of protoplast technology in plant improvement is to increase genetic variation. First, plants regenerated from protoplasts may exhibit somaclonal variation for agronomically useful characters. Second, by fusion of protoplasts it is possible to effect organelle transfer (chloroplasts and mitochondria) and gene transfer between sexually incompatible parents. Finally, protoplasts are useful in biotechnology for genetic transformation by direct uptake of foreign DNA or through procedures such as electroporation. However, the first step in applying protoplast technology to grapevine improvement is the availability of methods for plant regeneration from protoplasts. As indicated above, this has yet to be achieved but some progress has been made with isolation techniques.

GENE INSERTION

A prospect of biotechnology is that it may be possible to insert foreign genes into the genomes of traditional cultivars such as Cabernet Sauvignon and Chardonnay without altering the *genes* concerned with any of their other characteristics, including wine quality. Of special interest is the conferring of resistance to virus disease by incorporation of viral coat protein genes (Abel *et al.*, 1986; Beachy *et al.*, 1988; Cuozzo *et al.*, 1988) or by expression of virus satellite RNA (Harrison *et al.*, 1987). Another interesting possibility is the conferring of resistance to lepidopteran pests by incorporation into the grapevine genome of genes encoding production of *Bacillus thuringiensis* toxin (Barton *et al.*, 1987).

The current situation with the application of biotechnology to grapevine improvement is similar to that with protoplasts. There is much interest in the potential of biotechnology for genetic improvement of woody plants but the first step, the production of genetically transformed grapevines that express a marker gene, has yet to be reported.

There have been encouraging preliminary results in several laboratories, both with *Agrobacterium*-mediated transformation and with particle acceleration, but no genetically transformed grapevines have emerged. This is in contrast to other horticultural crops such as apple (James, 1987), and walnut (McGranahan *et al.*, 1988) where genetically transformed plants expressing marker genes have been reported. It is frustrat-

224 GENETIC IMPROVEMENT

ing that the grapevine should prove to be such recalcitrant material for transformation, but this technical blockage is likely to be temporary. Sustained investment in research is needed if the exciting possibilities of biotechnology are to become realities in viticulture.

Recommended reading

Einset, J. and Pratt, C. 1975. Grapes. In *Advances in fruit breeding* (ed. J. Janick and J.N. Moore), pp. 130–53. Purdue University Press, West Lafayette, Indiana.
Moore, J.N. and Janick, J. 1983. *Methods in fruit breeding.* Purdue University Press, West Lafayette, Indiana. 464 pp.
Olmo, H.P. 1976. Grapes. In *Evolution of crop plants* (ed. N.W. Simmonds), pp. 294–8. Longman, London.
Rives, M. 1971. Génétique et amélioration de la vigne. In *Traité d'ampélologie: science et techniques de la Vigne.* (ed. J. Ribéreau-Gayon and E. Peynaud). vol. 1 (*Biologie de la vigne, sol de vignobles*), pp. 171–219. Dunod, Paris.
Simmonds, N.W. 1979. *Principles of crop improvement.* Longman, London and New York. pp. 162–70.

Literature cited

Abel, P.P., Nelson, R.S., De, B., Hoffmann, N., Rogers, S.G., Fraley, R.T. and Beachy, R.N. 1986. Delay of disease development in transgenic plants that express the tobacco mosaic virus coat protein gene. *Science* **232**: 738–43.
Alleweldt, G. and Dettweiler, E. 1986. Ampelographic studies to characterize grapevine varieties. *Vigne Vini (Riv. Italiana Vitic. Enol.)* Suppl. **12**: 56–9.
Alleweldt, G. and Possingham, J.V. 1988. Progress in grapevine breeding. *Theor. Appl. Genet.* **75**: 669–73.
Antcliff, A.J. 1973. Comparison of some local and imported clones of important wine grape varieties. *Australian Grapegrower and Winemaker* **113**: 3–4.
Antcliff, A.J. 1975. Four new grape varieties released for testing *J. Aust. Inst. Agric. Sci.* **41**: 262–4.
Antcliff, A.J. 1980. Inheritance of sex in *Vitis. Ann. Amélior. Plantes* **30**: 113–22.
Antcliff, A.J., Newman, H.P. and Barrett, H.C. 1983. Variation in chloride accumulation in some American species of grapevine. *Vitis* **22**: 357–62.
Barlass, M. 1986. Development of methods applicable to the selection of mutants from *in vitro* culture of grapevines. In *Proc. Int. Symp. Nuclear technology and in vitro culture for plant improvement,* pp. 259–65. IAEA and FAO, Vienna.
Barlass, M. and Skene, K.G.M. 1978. *In vitro* propagation of grapevine (*Vitis vinifera*) from fragmented shoot apices. *Vitis* **17**: 335–40.
Barton, K.A., Whiteley, H.R. and Yang, N.S. 1987. *Bacillus thuringiensis* delta-endotoxin expressed in transgenic *Nicotiana tabacum* provides resistance to Lepidopteran insects. *Plant Physiology* **85**: 1103–9.
Beachy, R.N., Nelson, R.S., Register, J., Fraley, R.T. and Tumer, N. 1988. Transformation to produce virus-resistant plants. In *Genetic Improvements of agriculturally important crops: progress and issues* (Ed. R.T. Fraley, N.M. Frey, and J. Schell), pp. 47–53. Current Communications in Molecular Biology. Cold Spring Harbor Laboratory.
Becker, H. 1977. Methods and results of clonal selection in viticulture. *Acta Hort.* **75**: 111–12.

Becker, H. 1988. Boerner: The first rootstock immune to all phylloxera biotypes. *Proc. 2nd Int. Cool Climate Vit. & Oenol. Symp., Auckland, New Zealand, January 1988* (ed. R.E. Smart, R.J. Thornton, S.B. Rodriguez and J.E. Young), pp. 51–2. New Zealand Society for Viticulture and Oenology, Auckland.

Benin, A., Gasquez, J, Mahfoudi, A. and Bessis, R. 1988. Caractérisation biochimique des cépages de *Vitis vinifera* L. par électrophorèse d'isoenzymes foliaires: Essai de classification des variétés. *Vitis* 27: 157–72.

Bernstein, L., Ehlig, C.F. and Clark, R.A. 1969. Effect of grape rootstocks on chloride accumulation in leaves. *J. Amer. Soc. Hort. Sci.* 94: 584–90.

Boubals, D. 1959. Contribution à l'étude des causes de la résistance des Vitacées au mildiou de la vigne [*Plasmopara viticola* (B. & C.) Berl. & Det.] et leur mode de transmission héréditaire. *Ann. Amélior. Plantes* 9: 5–233.

Boubals, D. 1961. Etude des causes de la résistance des Vitacées a l'oidium de la vigne [*Uncinula necator* (Schw.) Burr.] et leur mode de transmission héréditaire. *Ann. Amélior. Plantes* 11: 401–500.

Boubals, D. and Pistre, R. 1978. Résistance de certaines Vitacées et des porte-greffes usuels en viticulture au nématode *Xiphinema index* et à l'inoculation par le virus du court-noué (GFV). In *Proc. Int. Symp. Génétique et Amélioration de la Vigne, Bordeaux*, pp. 199–207. INRA, Paris.

Bouquet, A. 1977. Amélioration génétique de la vigne: essai de définition d'un schéma de sélection applicable à la création de nouvelles variétés. *Ann. Amélior. Plantes* 27: 75–86.

Bouquet, A. 1981. Resistance to grape fanleaf virus in muscadine grape inoculated with *Xiphinema index*. *Pl. Dis.* 65: 791–3.

Bouquet, A. 1982. Premières observations sur le déterminisme génétique de la polyembryonie spontanée chez un hybride interspécifique *Vitis vinifera* × *V. riparia*. *Vitis* 21: 33–9.

Bouquet, A. 1983. Etude de la résistance au phylloxera radicicole des hybrides *Vitis vinifera* × *Muscadinia rotundifolia*. *Vitis* 22: 311–23.

Bouquet, A. 1986. Introduction dans l'espèce *Vitis vinifera* d'un caractère de résistance à l'oidium (*Uncinula necator* SCHW. BURR.) issu de l'espèce *Muscadinia rotundifolia* (MICHX.) SMALL. *VigneVini* (*Riv. Ital. Vitic. Enol.*) Suppl. 12: 141–6.

Bouquet, A. and Dalmasso, A. 1976. Comportement d'une collection de porte-greffes de vigne en présence d'une population de nématodes (*Meloidogyne* sp.) originaire du sud-ouest de la France. *Conn. Vigne Vin* 10: 161–74.

Bouquet, A. and Danglot, Y. 1983. Recherche de porte-greffes de vigne résistant à la transmission du virus court-noué (GFV) par le nématode *Xiphinema index* THORNE and ALLEN. 1. Application de la méthode ELISA à la réalisation d'un test rapide de sélection. *Agronomie* 3: 957–963.

Branas, J. 1980. Sol, vigne, qualité des vins. *Progr. Agric. Vitic.* 24: 529–532.

Cain, D.W., Emershad, R.L. and Tarailo, R.E. 1983. *In ovulo* embryo culture and seedling development of seeded and seedless grapes. *Vitis* 22: 9–14.

Cancellier, S. and Cossio, F. 1988. Risultati di osservazione su un clone di 'Corvina Veronese' (*Vitis vinifera* L.) moltiplicato attraverso la coltura *in vitro*. *Rivista di Viticoltura e di Enologia, Cogneliano* 41: 110–17.

Carbonneau, A. 1985. The early selection of grapevine rootstocks for resistance to drought conditions. *Am. J. Enol. Vitic.* 36: 195–8.

Cirami, R., McCarthy, M. and Furkaliev, D.G.J. 1985. Clonal selection and comparison in South Australia. *Australian Grapegrower and Winemaker* 262: 18–19.

Cuozzo, M., O'Connell, K.M., Kaniewski, W., Fang, R.X., Chua, N.H. and Tumer, N. 1988. Viral protection in transgenic tobacco plants expressing the cucumber mosaic virus coat protein or its antisense RNA. *Bio/Technology* **6**: 549–57.

Dal Belin Peruffo, A., Varanni, Z. and Maggioni, A. 1981. Carratterizzazione di specie, varieta e cloni di vite mediante elettrofocalizzazione di estratti enzimatici fogliari. *C.R.3 Symp. Intern. Selection Clonale de la Vigne, Venice, June 8–12, 1981*, pp. 31–40. CNR, Venice.

Davidis, U.X. and Olmo, H.P. 1964. *Vitis vinifera* × *V. rotundifolia* hybrids as phylloxera resistant rootstocks. *Vitis* **4**: 1329–143.

De Wald, S.G. and Moore, G.A. 1987. Somaclonal variation as a tool for the improvement of perennial fruit crops. *Fruit Var. J.* **41**: 54–7.

Duran-Vila, N., Juarez, J. and Arregui, J.M. 1988. Production of viroid-free grapevines by shoot tip culture. *Am. J. Enol. Vitic.* **39**: 217–20.

Favre, J.M. 1976. Influence de l'état physiologique de la plante et de la nature de l'organe prélevé sur l'obtention des néoformations caulinaires de la vigne. *Congr. Nat. Soc. Savantes Lille* **101** (1): 465–74.

Firoozabady, E. and Olmo, H.P. 1982. The heritability of resistance to rootknot nematode (*Meloidogyne incognita acrita* CHIT.) in *V. vinifera* × *V. rotundifolia* hybrid derivatives. *Vitis* **21**: 136–44.

Flores, R., Duran-Vila, N., Pallas, V. and Semancik, J. 1985. Detection of viroid and viroid-like RNAs from grapevine. *J. Gen. Virol.* **66**: 2095–102.

Fregoni, M. 1977. Effects of the soil and water on the quality of the harvest. *O.I.V. International Symposium on the Quality of the Vintage, Capetown, South Africa, February 14–21, 1977*, pp. 151–65. OVRI, Stellenbosch, South Africa.

Fregoni, M., Scienza, A. and Miravelle, R. 1978. Evaluation précoce de la résistance des porte-greffes a la sécheresse. *Génétique et Amélioration de la Vigne*, pp. 287–96. INRA, Paris.

Galet, P. 1979. *A practical ampelography* (transl. L.T. Morton). Cornell University Press, Ithaca. 248 pp.

Galzy, R. 1964. Technique de la thermothérapie des viroses de la vigne. *Ann. Epiphyt.* **15**: 245–56.

Grenan, S. 1982. Quelques réflexions à propos de modifications morphogénétiques consécutives à la culture *in vitro* chez vigne (*Vitis vinifera* L.). *Ann. Sci. Nat. Bot. Paris* **4**: 135–46.

Grenan, S. 1984. Polymorphisme foliaire consécutif à la culture *in vitro* de *Vitis vinifera* L. *Vitis* **23**: 159–74.

Harris, A.R. 1983. Resistance of some *Vitis* rootstocks to *Xiphinema index*. *J. Nematology* **15**: 405–9.

Harrison, B.D., Mayo, M.A. and Baulcombe, D.C. 1987. Virus resistance in transgenic plants that express cucumber mosaic virus satellite RNA. *Nature* **328**: 799–802.

Helentjaris, T. 1988. Use of RFLP analysis to identify genes involved in complex traits of agronomic importance. In *Communication in molecular biology* (ed. R.T. Fraley, N.M. Frey and J. Schell), pp. 27–30. Cold Spring Harbor Laboratory, New York.

Helentjaris, T., King, G., Slocum, M., Siedenstrang, C. and Wegman, S. 1985. Restriction fragment polymorphisms as probes for plant diversity and their development as tools for applied plant breeding. *Plant Molecular Biol.* **5**: 109–18.

Hirabayashi, T., Kozaki. I. and Akihama, T. 1976. *In vitro* differentiation of shoot from anther callus in *Vitis*. *HortScience* **11**: 511–12.

James, D.J. 1987. Cell and tissue culture technology for the genetic manipulation of temperate fruit trees. *Biotechnology and Genetic Engineering Reviews* 5: 33–79.

Jona, R. and Webb, K.J. 1978. Callus and axillary bud culture of *Vitis vinifera* 'Sylvaner riesling.' *Sci. Hort.* 9: 55–60.

Kim, S.K, Reisch, B.I. and Aldwinckle, H.S. 1986. *In vitro* grape shoot tip mutagenesis. *Vigne Vini (Riv. Italiana Vitic. Enol.)* Suppl. 12: 26–7.

Kissler, J.J. and Carlton, A B. 1969. The potential of wine grape production in the San Joaquin Delta area of California. *Am. J. Enol. Vitic.* 20: 40–7.

Klempka, K.C., Meredith, C.P. and Sall, M.A. 1984. Dual culture of grape powdery mildew (*Uncinula necator* BURR.) on its host (*Vitis vinifera* L.). *Am. J. Enol. Vitic.* 35: 170–4.

Koltunow, A.M. and Rezaian, M.A. 1988. Grapevine yellow speckle viroid: structural features of a new viroid group. *Nucleic Acids Research* 16: 849–64.

Kunde, R.M., Lider, L.A. and Schmitt, R.V. 1968. A test of *Vitis* resistance to *Xiphinema index. Am. J. Enol. Vitic.* 19: 30–6.

Langcake, P. and Pryce, R.J. 1976. The production of resveratrol by *Vitis vinifera* and other members of the *Vitaceae* as a response to infection or injury. *Physiol. Plant Pathol.* 9: 77–86.

Larkin, P.J. and Scowcroft, W.R. 1981. Somaclonal variation: a novel source of variability from cell cultures for plant improvement. *Theor. Appl. Genet.* 60: 197–214.

Lebrun, L., Rajasekaran, K. and Mullins, M.G. 1985. Selection *in vitro* for NaCl-tolerance in *Vitis rupestris* SCHEELE. *Ann. Bot.* 56: 733–9.

Lee, T.C. and Wicks, T. 1982. Dual culture of *Plasmopara viticola* in grapevine and its application to systemic fungicide evaluation. *Plant Dis.* 66: 308–10.

Lefort, P.L. and Leglise, N. 1977. Quantitative stock-scion relationships in vine: Preliminary investigations by the analysis of reciprocal graftings. *Vitis* 16: 149–61.

Lider, L.A. 1954. Inheritance of resistance to a rootknot nematode (*Meloidogyne incognita* var. *acrita* CHITWOOD) in *Vitis* spp. *Proc. Helminthol. Soc. Wash. D.C.* 21: 53–60.

Lider, L.A. and Goheen, A.C. 1986. Field resistance to the grapevine fanleaf virus–*Xiphinema index* complex in interspecific hybrids of *Vitis. Vigne Vini (Riv. Ital. Vitic. Enol.)* Suppl. 12: 166–69.

Mauro, M.C., Vaillant, V., Tey-Rulh, P., Mathieu, Y. and Fallot, J. 1988. *In vitro* study of the relationship between *Vitis vinifera* and *Eutypa lata* (PERS. FR.) TUL. I. Demonstration of the toxic compounds secreted by the fungus. *Am. J. Enol. Vitic.* 39: 200–4.

McGranahan, G.H., Leslie, C.A., Uratsu, S.L., Martin, L.A. and Dandekar, A.M. 1988. *Agrobacterium*-mediated transformation of walnut somatic embryos and regeneration of transgenic plants. *Bio/Technology* 6: 800–4.

Meredith, C.P., Lider, L.A., Raski, D.J and Farari, N.L. 1982. Inheritance of tolerance to *Xiphinema index* in *Vitis* species. *Am. J. Enol. Vitic.* 33: 154–8.

Morel, G. 1944. Le développement de mildiou sur des tissus de vigne cultivés *in vitro. C.R. Hebd. Séances Acad. Sci.* 218: 50–2.

Morel, G. 1948. Cultures associées de tissus végétaux et de parasites obligatoires. *Ann. Epiphyt.* 14: 78–84.

Mullins, M.G., Nair, Y. and Sampet. P. 1979. Rejuvenation *in vitro*: induction of juvenile characters in an adult clone of *Vitis vinifera* L. *Ann Bot.* 44: 623–7.

Mullins, M.G. and Srinivasan, C. 1976. Somatic embryos and plantlets from an ancient clone of the grapevine (cv. Cabernet sauvignon) by apomixis *in vitro*. *J. Exp. Bot.* **27**: 1022–30.

Newman, H.P. and Antcliff, A.J. 1983. Chloride accumulation in some hybrids and backcrosses of *Vitis berlandieri* and *Vitis vinifera*. *Vitis* **23**: 106–12.

Olmo, H.P. 1948. Ruby Cabernet and Emerald riesling. Two new table-wine grape varieties. *Calif. Agr. Expt. Sta. Bull.* no. 404.

Parfitt, D.E. and Arulsekar, S. 1989. Inheritance and isozyme diversity for GPI and GPM among grape cultivars. *J. Am. Soc. Hort. Sci.* **114**: 486–91.

Pena-Iglesias, A. and Vecino, B. 1987. Cytological studies of grapevine leafroll infected tissue: Further evidence of viroid etiology and improvement of diagnosis. *Vitis* **26**: 37–41.

Pouget, R. 1977. Obtention de nouveaux porte-greffes favorables à la qualité: la résistance à la chlorose et la maitrise de la vigueur. *Bulletin de l'O.I.V.* **556**: 389–97.

Pouget, R. 1980. Breeding grapevine rootstocks for resistance to iron chlorosis. *Proc. 3rd Int. Symp. Grape Breeding, Davis, Calif.*, pp. 191–7. University of California, Davis.

Pouget, R and Delas, J. 1982. Interaction entre le greffon et le porte-greffe chez la vigne. Application de la méthode des greffages réciproques à l'étude de la nutrition minérale. *Agronomie* **2**: 231–42.

Pouget, R. and Ottenwaelter, M. 1973. Etude méthodologique de la résistance à la chlorose calcaire chez la vigne: principe de la méthode des greffages réciproques et application à la recherche de porte-greffes résistants. *Ann. Amélior. Plantes* **23**: 347–56.

Pouget, R. and Ottenwaelter, M. 1986. Recherches de porte-greffes adaptés aux sols acides: une nouvelle variété, le Gravesac. *Vigne Vini (Riv. Italiana Vitic. Enol.)* Suppl. **12**: 134–7.

Rajasekaran, K. and Mullins, M.G. 1983*a*. Influence of genotype and sex expression on formation of plantlets by cultured anthers of grapevines. *Agronomie* **3**: 233–8.

Rajasekaran, K. and Mullins, M.G. 1983*b*. The origin of embryos and plantlets from cultured anthers of hybrid grapevines. *Am. J. Enol. Vitic.* **34**: 108–13.

Ramming, D.W. and Emershad, R.L. 1982. *In ovulo* embryo culture of seeded and seedless *Vitis vinifera* L. *HortScience* **17**: 487.

Rankine, B.C., Fornachon, J.C.M., Boehm, E.W. and Cellier, K.M. 1971. Influence of grape variety, climate, and soil on grape composition and quality of table wines. *Vitis* **10**: 33–50.

Reisch, B.I., Aldwinckle, H.A. Roberts, M.H. and Kim, S.K 1985. I. Embryogenesis from petiole cultures of 'Horizon' grapes. II. *In vitro* grape shoot tip mutagenesis. *Colloque Amélioration de la Vigne et Culture in vitro, Paris, April 23–24, 1985*, pp. 161–170. Moët-Hennessy, Paris.

Rezaian, M.A., Koltunow, A.M. and Krake, L.R. 1988. Isolation of three viroids and a circular RNA from grapevines. *J. Gen. Virol.* **69**: 413–22.

Rives, M. 1961. Bases génétiques de la sélection clonale chez la vigne. *Ann. Amélior. Plantes* **11**: 337–48.

Saayman, D. 1977. The effect of soil and climate on wine quality. *O.I.V. International Symposium on the Quality of the Vintage, Cape Town, South Africa, February 14–21, 1977*, pp. 197–208. OVRI, Stellenbosch, South Africa.

Sano, T., Uyida, I., Shikata, E., Meshi, T., Ohno, T. and Okada, Y.N. 1985. A viroid-like RNA isolated from grapevine has high sequence homology with hop stunt viroid. *J. Gen. Virol.* **66**: 333–8.

Sauer, M.R. 1968. The effect of vine rootstocks on chloride concentration in Sultana scions. *Vitis* **7**: 223–6.

Semancik, J.S., Rivera-Bustamente, R. and Goheen, A.C. 1987. Widespread occurrence of viroid-like RNA's in grapevines. *Am. J. Enol. Vitic.* **38**: 35–40.

Skene, KG.M. 1974. Culture of protoplasts from grapevine pericarp callus. *Aust. J. Plant Physiol.* **1**: 371–6.

Skene, K.G.M. and Barlass, M. 1983. Studies of the fragmented shoot apex of grapevine. IV. Separation of phenotypes in a periclinal chimera *in vitro*. *J. Exp. Bot.* **34**: 1271–80.

Soller, M. and Beckmann, J.S. 1983. Genetic polymorphism in varietal identification and genetic improvement. *Theor. Appl. Genet.* **67**: 25–33.

Stein, U. and Hoos, G. 1984. Induktions- und Nachweismethoden fur Stilbene bei Vitaceen. *Vitis* **213**: 179–94.

Swanepoel, J.J. and De Villiers, C.E. 1987. A numerical-taxonomic classification of *Vitis* spp. and cultivars based on leaf characteristics. *S. Afr. J. Enol. Vitic.* **8**: 31–5.

Szychowski, J.A., Goheen, A.C. and Semancik, J.S. 1988. Mechanical transmission and rootstock reservoirs as factors in the widespread distribution of viroids in grapevines. *Am. J. Enol. Vitic.* **39**: 213–20.

Tanksley, S.D., Young, N.D., Patterson, A.H. and Bonierbale, M.W. 1989. RFLP mapping in plant breeding: New tools for an old science. *Bio/Technology* **7**: 257–64.

Valat, C. and Rives, M. 1973. Information and comments on variation induced by thermotherapy. *Riv. Pathol. Veg.* **4**: 291–3.

Vilaplana, M. and Mullins, M.G. 1989. Regeneration of grapevines (*Vitis* spp.) *in vitro*: Formation of adventitious buds on hypocotyls and cotyledons of somatic embryos. *J. Plant Physiology* **134**: 413–19.

Wagner, R. and Bronner, A. 1974. Etude de la fertilité des semis de *Vitis vinifera*: Application a la mise au point d'un test précoce de sélection. *Ann. Amélior. Plantes* **24**: 145–57.

Walker, M.A., Meredith, C.P. and Goheen, A C. 1985. Sources of resistance to grapevine fanleaf virus (GVF) in *Vitis* species. *Vitis* **24**: 218–28.

Wolfe, W.H. 1976. Identification of grape varieties by isozyme banding patterns. *Am. J. Enol. Vitic.* **27**: 68–73.

Zimmermann, H. and Becker, N.J. 1978. La sélection de porte-greffes à base de *Vitis cinerea*. *Proc. Int. Symp. Génétique et Amélioration de la Vigne, Bordeaux*, pp. 281–6. INRA, Paris.

Zou, C. and Li, P. 1981. Induction of pollen plants of grape (*Vitis vinifera* L.) *Acta Bot. Sinica* **23**: 79–81.

Index

abscisic acid (ABA) 86, 88, 95, 103, 120, 131, 133
abscission
  flower 122, 178
  leaves 59, 107, 156
Acreosperma 17
Acarina 181
adenosine triphosphate (ATP) 80, 81, 89
adjuvant 180
adventitious roots 60
Africa 19
Afuz-Ali 33
Agoston Haraszthy 9
Agrobacterium tumefaciens (*see also* crown gall) 188, 192
alcohol 138
Algeria 9, 10
Alicante Bouschet 31, 135
Allobrogica 32
aluminum 168
America 19
Ametadoria (*Sturmia*) *harrisinae* 186
amino acids 134, 140
ammonia 134, 140
Ampelidaceae 6, 17
Ampelideae 17
Ampelocissus 18, 19, 23, 26
ampelography 33, 34, 59, 60, 209
ampelometry 33
Ampelopsis 18, 19
amyloplasts 54
anaerobiosis 94
Anagrus epos 186
androecium 70
Anlage (pl. Anlagen) 61, 65
  regulation of development 112, 114, 115, 117
anthers 70, 220, 222
anthesis 66, 68, 96, 98, 99, 103, 163, 178
anthocyanidin 135

anthocyanins 125, 129, 130, 134, 140, 158
anthracnose 188
aoûtement 54, 65
Apanteles harrisinae 186
aperitif wines 9
Apianae 32
apical meristem 38, 41, 43, 50
apoplast 91, 101, 102
Appellation d'Origine Contrôlée (AOC) 204
archesporium 72
Argentina 8, 9, 27
arginine 106, 129, 134, 138, 140
Arizona 24
Armagnac 9, 11
Armillaria mellea (Armillaria root rot) 188
arms 54, 98
Asia 19
Australia 9, 14, 15, 27, 29, 33, 171, 183
Austria 33
auxin 122, 131
axillary bud 39, 46
AXR # 1 100, 195

Bacillus thuringiesis 188, 223
Bacterial blight 188
bacterial diseases 188, 191, 211
Barbera 29, 164
base temperature 92, 96
Beaujolais 204
benzyladenine 114
berry growth
  double sigmoid pattern 123, 125, 126
  Stage I 124, 128, 130, 131, 134, 139
  Stage II 125, 129, 131, 139
  Stage III 125–9, 131, 139, 140
biotechnology 205, 223
Biturica 32
Black Corinth (syn. Zante Currant) 11, 31, 178

231

malic enzyme 132, 133, 140
malvidin 135
manganese (Mn) 161, 163, 168
mass flow 103
medullary (rays) 53, 57
megaspores 70
Meloidogyne (root knot nematode) 184, 211
  M. arenaria 216, 217
  M. hapla 216
  M. incognita 216, 217
  M. javanica 216, 217
Melon 31
mesocarp 68, 76, 125
methyl anthranilate 137
methyl bromide 187
Meunier 31
Mexico 8, 25
microclimate 154, 158
microgametogenesis 73
micronutrients 160, 161, 165
micropyle 68, 69, 74, 75
microsporangia 72
microspores 73
microsporogenesis 72
mineral nutrition 120, 160
mineral nutrients 80, 104, 107, 165, 167–9
  reserves 106
mitochondria 89
monoglucosides 22, 135
monopodial theory 49
monopodium 43
monoterpenes 158
morphactins 180
morphology
  adult 38, 40
  juvenile 38, 40
Müller-Thurgau 176, 210
Muscadinia 17, 18, 23, 25–7, 44, 187, 210, 212
  M. munsoniana 25
  M. popenoei 25
  M. rotundifolia 25–7, 179, 211, 212, 216, 217
Muscat Gordo (syn. Muscat of Alexandria) 11
Muscat Hamburg 115
Muscat of Alexandria 1, 65, 117, 118, 164
Muscats 32, 137, 176

Nebbiolo 29
nectaries 68
nematodes (*see also Meloidogyne, Pratylenchus, Tylenchulus and Xiphinema*) 29, 181, 184, 186, 187
  sources of resistance 211, 215, 216
nepovirus 193, 194

New Mexico 24
New Zealand 9, 12, 183
nitrate 103, 134
nitrogen (N)
  concentrations in tissues 160, 163, 165, 166
  content in tissues 134, 138, 167, 168
  fertilization 106, 120, 138, 169
  functions 161
  reserves 104, 106
  transport 103
nobel rot 190
node 44
nodosities 181, 182
Noiriens 31
North America 8, 18
Nucellus 68, 69, 75
nucleoproteins 162

Odjalechi 31
Office International de la Vigne et du Vin (OIV) 10, 34
Ohanez 68, 117, 118
oidium: *see* powdery mildew
oleanolic acid 76
organic acids 124, 131, 133, 162
organogenesis 219, 221
Orthoptera 181
ovary 68
overcropping 159
ovules 68
oxidative pentose phosphate pathway 89
ozone 147

paclobutrazol 179
Paleovitis 22
palisade tissue 59
palmate venation 59
parthenocarpy 75, 122, 126
Parthenocissus 18, 19
pearls 43, 44
pedicel 68, 76, 77
pellicule 77
peonidin 135
pericarp 68, 76, 131
  growth of 122, 124, 125, 128
pericycle 57
periderm 54, 55, 60
Perlette 83
Peru 8, 12
petals 65, 66
petiole 44, 58, 59
  sinus 59
  assessing fertilization needs 163
petunidin 135
pH 138, 140, 157, 158, 162